Textbook of Marine Biology

Textbook of Marine Biology

Anthony Moore

Larsen & Keller
www.larsen-keller.com

Textbook of Marine Biology
Anthony Moore
ISBN: 978-1-64172-667-2 (Hardback)

Larsen & Keller

Published by Larsen and Keller Education,
5 Penn Plaza,
19th Floor,
New York, NY 10001, USA

Cataloging-in-Publication Data

Textbook of marine biology / Anthony Moore.
p. cm.
Includes bibliographical references and index.
ISBN 978-1-64172-667-2
1. Marine biology. 2. Aquatic biology. 3. Marine sciences. I. Moore, Anthony.
QH91 .T49 2022
578.77--dc23

For more information regarding Larsen and Keller Education and its products, please visit the publisher's website www.larsen-keller.com

Table of Contents

Preface VII

Chapter 1 **Introduction to Marine Biology** **1**
- a. Marine Microbiology 2
- b. Marine Life 3
- c. Marine Ecology 4
- d. Marine Habitat 8

Chapter 2 **Marine Fishes, Reptiles and Birds** **11**
- a. Marine Fishes 11
- b. Marine Reptiles 18
- c. Marine Birds 42

Chapter 3 **Marine Microorganisms** **75**
- a. Prokaryotes 78
- b. Eukaryotes 85

Chapter 4 **Marine Mammals** **101**
- a. Cetacean 105
- b. Pinnipeds 123
- c. Mustelidae 126
- d. Sirenians 128
- e. Polar Bears 130

Chapter 5 **Marine Ecosystem** **132**
- a. Coral Reef Ecosystem 145
- b. Mangrove Ecosystem 163
- c. Kelp Forest Ecosystem 169
- d. Rocky Shore Ecosystem 173
- e. Sandy Beach Ecosystem 179
- f. Salt Marsh Ecosystem 195
- g. Deep Sea Ecosystem 199

Permissions

Index

Preface

The branch of science which studies organisms in the seas and oceans is known as marine biology. Classification in this field is done usually on the basis of environment rather than on the taxonomy. The habitats in marine biology are primarily divided into coastal and open ocean habitats. Some of these habitats are seagrass meadows, coral reefs and kelp forests. There are different types of organisms that inhabit the ocean which are broadly classified into microscopic life, vertebrates, plants and algae, fungi and invertebrates. Microbes in the ocean are responsible for most of the photosynthesis which occurs in the ocean. Some of the vertebrates which are studied within this discipline are fish, marine mammals and seabirds. This book brings forth some of the most innovative concepts and elucidates the unexplored aspects of marine biology. It is an upcoming field of science that has undergone rapid development over the past few decades. This book will provide comprehensive knowledge to the readers.

To facilitate a deeper understanding of the contents of this book a short introduction of every chapter is written below:

Chapter 1- The domain of biology which is involved in the study of marine life is called marine biology. Some of the interest areas within this field of study are marine life, marine ecology and marine habitat. This is an introductory chapter which will provide a brief introduction to these significant aspects of marine biology.

Chapter 2- There are a number of different vertebrate which are studied within marine biology such as fishes, reptiles and birds. Some of the marine reptiles studied within this area of study are sea turtle, sea snake, marine iguana and crocodile. The chapter closely examines these marine animals to provide an extensive understanding of the subject.

Chapter 3- The marine microorganisms are broadly divided into two categories, namely, prokaryotes and eukaryotes. Marine prokaryotes are further subdivided into marine viruses, marine bacteria and marine archaea. The chapter closely examines these key categories of marine microorganisms to provide an extensive understanding of the subject.

Chapter 4- The aquatic mammals which rely on marine ecosystems such as the ocean for their sustenance and existence are termed as marine mammals. There are a number of categories of marine mammals such as Cetaceans, Pinnipeds, Mustelidae and Sirenians. This chapter has been carefully written to provide an easy understanding of the varied facets of these types of marine mammals.

Chapter 5- The primary characteristic of a marine ecosystem is water that has a high salt content. There are a number of marine ecosystems such as coral reef ecosystem, mangrove ecosystem, kelp forest ecosystems and rocky shore ecosystem. The chapter closely examines these key types of marine ecosystems to provide an extensive understanding of the subject.

I would like to share the credit of this book with my editorial team who worked tirelessly on this book. I owe the completion of this book to the never-ending support of my family, who supported me throughout the project.

Anthony Moore

1

Introduction to Marine Biology

The domain of biology which is involved in the study of marine life is called marine biology. Some of the interest areas within this field of study are marine life, marine ecology and marine habitat. This is an introductory chapter which will provide a brief introduction to these significant aspects of marine biology.

Marine biology is the science that deals with animals and plants that live in the sea. It also deals with air-borne and terrestrial organisms that depend directly upon bodies of salt water for food and other necessities of life. In the broadest sense it attempts to describe all vital phenomena pertaining to the myriads of living things that dwell in the vast oceans of the world. Some of its specialized branches concern natural history, taxonomy, embryology, morphology, physiology, ecology, and geographical distribution. Marine biology is closely related to the science of oceanography because of the relationship of the physical features of the oceans to the living organisms that dwell in them. It aids in the understanding of marine geology through the study of those organisms that contribute their skeletal remains to the floors of the oceans or that elaborate the vast coral reefs of the tropic seas.

A principal aim of marine biology is to discover how ocean phenomena control the distribution of organisms. Marine biologists study the way in which particular organisms are adapted to the various chemical and physical properties of the seawater, to the movements and currents of the ocean, to the availability of light at various depths, and to the solid surfaces that make up the seafloor. Special attention is given to determining the dynamics of marine ecosystems, particularly to the understanding of food chains and predator-prey relationships. Marine biological information on the distribution of fish and crustacean populations is of great importance to fisheries. Marine biology is also concerned with the effects of certain forms of pollution on the fish and plant life of the oceans, particularly the effects of pesticide and fertilizer runoff from land sources, accidental spills from oil tankers, and silting from coastline construction activities.

During the second half of the 19th century, when the emphasis was on the collection, description, and cataloging of marine organisms, methods evolved for the capture and preservation of specimens for study. Marine biologists adapted traditional dredges and trawls to collect specimens from the ocean floor; and hoop nets were used to secure free-swimming animals. New instruments for collecting water samples and obtaining temperature information at any desired depth were developed.

Late in the 19th century, the focus began to shift from collecting and cataloging to the systematic analysis of marine ecosystems and the ecological roles and behaviour of marine life. By the early 20th century, oceanographers had begun to intensively study fishing grounds and other localities of economic importance. This research combined studies of marine flora and fauna, ocean currents, water temperature, salinity, and oxygen levels, and other factors in an effort to understand the relationship between marine animals and their environment.

Since World War II, direct observation of marine organisms in their natural habitats has been made possible by underwater cameras, television, improved diving equipment, and submersible craft, or submarines that can descend to great depths. Underwater television provides the observer with a continuous picture of events that occur within the field of the submerged camera. The development of self-contained diving equipment made it possible for the investigator to inspect marine organisms in their natural habitat.

Morphological and taxonomic studies of marine organisms are generally performed on preserved materials in connection with the work in museums and universities. Physiological and embryological investigations requiring the use of living material are generally pursued at biological stations. These are situated on the seacoast, thus facilitating the rapid transfer of specimens to the laboratory where they may be maintained in seawater provided by special circulating systems.

Marine Microbiology

Marine microbiology is the study of microorganisms and non-organismic microbes that exist in saltwater environments, including the open ocean, coastal waters, and estuaries, on marine surfaces and in sediments. It focuses strongly on interaction within communities of microorganisms and between these communities and macroorganisms, both within the ocean and on the rest of the planet.

Complicated Identities

Defining marine microbiology means determining what exactly qualifies as a marine microorganism. This isn't a simple proposition; biology overall has been confounded by how to classify them since microscopic beings were first discovered. "Microorganism" usually means any living entity too small to see with the naked eye and is often used for any single-celled organism. However, some single-celled organisms are visible alone or when gathered in communities and some multicellular animals are microscopic. Additionally, some microbes, viruses, virions, prions, and viroids, are not actually alive by current definitions. These are the "non-organismic microbes." Overall, marine microbiology deals with all very small life and life-like biological phenomena: non-organismic microbes, bacteria, archaea, protozoans, single-celled algae and very small multicellular plants, fungi, and animals.

Multiple Locations

Marine microbiology isn't just for the open ocean. It's also concerned with microbial communities in coastal waters and estuaries, where saltwater meets fresh. Marine microorganisms are additionally found on maritime surfaces, such as sea cliffs splashed with ocean spray, and in sediments: those underneath the ocean as well as on coastlines and in estuaries, and those left farther inland by retreat of ocean waters.

Flowing through the Food Web

Marine microorganisms rarely exist alone. Rather, they combine into communities, where they often depend on one another for food. These communities join with other small lifeforms, including the larvae of many invertebrates and fish, to form the enormous living waves of plankton that marine macroorganisms, from tiny filter-feeding invertebrates to gigantic whales, depend on for food. While often classified as zooplankton (animal-like floaters) or phytoplankton (plant-like floaters), many marine microorganisms are mixotrophic, which means they can behave either like plants or like animals by switching between photosynthesis on the one hand, and devouring other microorganisms on the other.

Supporting Land-locked Life

Dependence on marine microorganisms does not stop where sea meets land. These diminutive creatures support all life on earth. They include bacteria that cycle nitrogen and the oxygen-cycling algae, which generate at least 50 percent Earth's oxygen. They are part of the global nutrient flow that keeps sulfur, phosphorous and carbon circulating through the ecosphere. Studying their interdependence with one another also gives scientists insight into the phenomenon of evolution and may provide clues to the relationships that made multicellular organisms and land-based life possible in the first place.

Marine Life

The phrase 'marine life' refers to organisms that live in salt water. These can include a diverse array of plants, animals and microbes (tiny organisms) such as bacteria and archaea.

Marine Life are Adapted to Life in Saltwater

From the perspective of a land animal like us, the ocean can be a harsh environment. However, marine life is adapted to live in the ocean. Characteristics that help marine life thrive in a saltwater environment include the ability to regulate their salt intake or deal with large quantities of salt water, adaptations to obtain oxygen (e.g., a fish's gills), being able to withstand high water pressures, living in a place where they can get enough light, or being able to adjust to a lack of light. Animals and plants that live on the edge of the ocean, such as tide pool animals and plants, also need to deal with extremes in water temperature, sunlight, wind and waves.

Types of Marine Life

There is a huge diversity in marine species. Marine life can range from tiny, single-celled organisms to gigantic blue whales, which are the largest creatures on Earth. Below is a list of the major phyla, or taxonomic groups, of marine life.

Major Marine Phyla

The classification of marine organisms is always in flux. As scientists discover new species, learn more about the genetic makeup of organisms, and study museum specimens, they debate how organisms should be grouped. More information about the major groups of marine animals and plants is listed below.

Marine Animal Phyla

Some of the most well-known marine phyla are listed below. You can find a more complete list here. The marine phyla listed below are drawn from the list on the World Register of Marine Species.

- Annelida - This phylum contains segmented worms. An example of a segmented marine worm is the Christmas tree worm.
- Arthropoda - Arthropods have a segmented body, jointed legs and a hard exoskeleton for protection. This group includes lobsters and crabs.
- Chordata - Humans are in this phylum, which also includes marine mammals (cetaceans, pinnipeds, sirenians, sea otters, polar bears), fish, tunicates, seabirds and reptiles.
- Cnidaria - This is a diverse phylum of animals, many of whom have stinging structures called nematocysts. Animals in this phylum include corals, jellyfish, sea anemones, sea pens and hydras.
- Ctenophora - These are jelly-like animals, such as comb jellies, but they don't have stinging cells.
- Echinodermata - This is one of my favorite phylums. It includes such beautiful animals as sea stars, brittle stars, basket stars, sand dollars and sea urchins.
- Mollusca - This phylum includes snails, sea slugs, octopuses, squids, and bivalves such as clams, mussels and oysters.
- Porifera - This phylum includes sponges, which are living animals. They can be very colorful and come in a diverse array of shapes and sizes.

Marine Plant Phyla

There are also several phyla of marine plants. These include the Chlorophyta, or green algae, and the Rhodophyta, or red algae.

Marine Ecology

Marine Ecology is the scientific study of marine-life habitat, populations, and interactions among organisms and the surrounding environment including their abiotic (non-living physical and chemical factors that affect the ability of organisms to survive and reproduce) and biotic factors (living things or the materials that directly or indirectly affect an organism in its environment).

Marine ecology is a subset of the study of marine biology and includes observations at the biochemical, cellular, individual, and community levels as well as the study of marine ecosystems and the biosphere.

The study of marine ecology also includes the influence of geology, geography, meteorology, pedology, chemistry, and physics on marine environments. The impact of human activity such as medical research, development, agriculture, fisheries, and forestry is also studied under marine ecology. In some ways, marine ecology is more complex than the relatively straightforward study of a particular organism or environment because of the numerous interconnections, symbiotic relationships, and influence of many factors on a particular environment.

To understand the difference between marine biology and marine ecology, it may be useful to look at a community of organisms. A marine biologist may focus on behavioral relationships between the organisms in one particular species while someone studying ecology would study how the behavior of one organism influences another. An ecologist would also look at abiotic factors and how they influence that organism. A scientist studying community ecology might study a group of organisms to see how they influence other species and abiotic factors.

The major subcategories of ecology are:

- Physiological ecology: The study of how biotic and abiotic factors act on the physiological characteristics of an organism and how the organism adapts to the abiotic and biotic environment.
- Behavioral ecology: A subcategory of ecology that studies which ecological and evolutionary dynamics is responsible for the way in which organisms adapt to their environment.
- Population ecology: The study of populations of organisms in a particular species and how the populations interact with their environment.
- Community ecology: The study of how species react to each other in a community.
- Landscape ecology: The study of how organisms interact with a particular landscape.
- Ecosystem ecology: The study of how energy and matter flow through ecosystems.
- Global ecology: The study of how energy and matter interact in the entire web of life on Earth.

The study of ecology in general includes all of the subcategories listed above as they apply to marine ecology, animal ecology, plant ecology, insect ecology, arctic ecology, tropical ecology and desert ecology.

Marine Environments

Classification of marine environments for ecological purposes is based very generally on two criteria, the dominant community or ecosystem type and the physical-geological setting. Those ecosystems identified by their dominant community type include mangrove forests, coastal salt marshes, submersed sea grasses and seaweeds, and tropical coral reefs. Marine environments identified by their physical-geological setting include estuaries, coastal marine and nearshore zones, and open-ocean-deep-sea regions.

An estuary is a semi enclosed area or basin with an open outlet to the sea where fresh water from the land mixes with seawater. The ecological consequences of fresh-water input and mixing create strong gradients in physical-chemical characteristics, biological activity and diversity, and the potential for major adverse impacts associated with human activities. Because of the physical forces of tides, wind, waves, and fresh-water input, estuaries are perhaps the most ecologically complex marine environment. They are also the most productive of all marine ecosystems on an area basis and contain within their physical boundaries many of the principal marine ecosystems defined by community type.

Coastal and near shore marine ecosystems are generally considered to be marine environments bounded by the coastal land margin (seashore) and the continental shelf 300–600 ft (100–200 m) below sea level. The continental shelf, which occupies the greater area of the two and varies in width from a few to several hundred kilometers, is strongly influenced by physical oceanographic processes that govern general patterns of circulation and the energy associated with waves and currents. Ecologically, the coastal and near shore zones grade from shallow water depths, influenced by the adjacent landmass and input from coastal rivers and estuaries, to the continental shelf break, where oceanic processes predominate. Biological productivity and species diversity and abundance tend to decrease in an offshore direction as the food web becomes supported only by planktonic production. Among the unique marine ecosystems associated with coastal and nearshore water bodies are seaweed-dominated communities (for example, kelp "forests"), coral reefs, and upwellings.

Approximately 70% of the Earth's surface is covered by oceans, and more than 80% of the ocean's surface overlies water depths greater than 600 ft (200 m), making open-ocean–deep-sea environments the largest, yet the least ecologically studied and understood, of all marine environments. The major oceans of the world differ in their extent of landmass influence, circulation patterns, and other physical-chemical properties. Other major water bodies included in open-ocean–deep-sea environments are the areas of the oceans that are referred to as seas. A sea is a water body that is smaller than an ocean and has unique physical oceanographic features defined by basin morphology. Because of their circulation patterns and geomorphology, seas are more strongly influenced by the continental landmass and island chain structures than are oceanic environments.

Within the major oceans, as well as seas, various oceanographic environments can be defined. A simple classification would include water column depths receiving sufficient light to support photosynthesis (photic zone); water depths at which light penetration cannot support photosynthesis and which for all ecological purposes are without light (aphotic zone); and the benthos or bottom-dwelling organisms. Classical oceanography defines four depth zones; epipelagic, 0–450 ft (0–150 m), which is variable; mesopelagic, 450–3000 ft (150–1000 m); bathypelagic, 3000–12,000 ft (1000– 4000 m); and abyssopelagic, greater than 12,000 ft (4000 m). These depth strata correspond approximately to the depth of sufficient light penetration to support photosynthesis; the zone in which all light is attenuated; the truly aphotic zone; and the deepest oceanic environments.

Marine Ecological Processes

Fundamental to marine ecology is the discovery and understanding of the principles that underlie the organization of marine communities and govern their behavior, such as controls on population growth and stability, quantifying interactions among populations that lead to persistent communities, and coupling of communities to form viable ecosystems. The basis of this organization is the flow of energy and cycling of materials, beginning with the capture of radiant solar energy through the processes of photosynthesis and ending with the remineralization of organic matter and nutrients.

Photosynthesis in seawater is carried out by various marine organisms that range in size from the microscopic, single-celled marine algae to multicellular vascular plants. The rate of photosynthesis, and thus the growth and primary production of marine plants, is dependent on a number of factors, the more important of which are availability and uptake of nutrients, temperature, and intensity and quality of light. Of these three, the last probably is the single most important in governing primary production and the distribution and abundance of marine plants. Considering the high attenuation of light in water and the relationships between light intensity and photosynthesis, net autotrophic production is confined to relatively shallow water depths. The major primary producers in marine environments are intertidal salt marshes and mangroves, submersed seagrasses and seaweeds, phytoplankton, benthic and attached microalgae, and—for coral reefs—symbiotic algae (zooxanthellae). On an aerial basis, estuaries and nearshore marine ecosystems have the highest annual rates of primary production. From a global perspective, the open oceans are the greatest contributors to total marine primary production because of their overwhelming size.

The two other principal factors that influence photosynthesis and primary production are temperature and nutrient supply. Temperature affects the rate of metabolic reactions, and marine plants show specific optima and tolerance ranges relative to photosynthesis. Nutrients, particularly nitrogen, phosphorus, and silica, are essential for marine plants and influence both the rate of photosynthesis and plant growth. For many phytoplankton-based marine ecosystems, dissolved inorganic nitrogen is considered the principal limiting nutrient for autotrophic production, both in its limiting behavior and in its role in the eutrophication of estuarine and coastal waters.

Marine food webs and the processes leading to secondary production of marine populations can be divided into plankton-based and detritus-based food webs. They approximate phytoplankton-based systems and macrophyte-based systems. For planktonic food webs, current evidence suggests that primary production is partitioned among groups of variously sized organisms, with small organisms, such as cyanobacteria, playing an equal if not dominant role at times in aquatic productivity. The smaller autotrophs—both through excretion of dissolved organic compounds to provide a substrate for bacterial growth and by direct grazing by protozoa (microflagellates and ciliates)—create a microbially based food web in aquatic ecosystems, the major portion of autotrophic production and secondary utilization in marine food webs may be controlled, not by the larger organisms typically described as supporting marine food webs, but by microscopic populations.

Macrophyte-based food webs, such as those associated with salt marsh, mangrove, and seagrass ecosystems, are not supported by direct grazing of the dominant vascular plant but by the production of detrital matter through plant mortality. The classic example is the detritus-based food webs of coastal salt marsh ecosystems. These ecosystems, which have very high rates of primary production, enter the marine food web as decomposed and fragmented particulate organics. The particulate organics of vascular plant origin support a diverse microbial community that includes bacteria, flagellates, ciliates, and other protozoa. These organisms in turn support higher-level consumers.

Both pelagic (water column) and benthic food webs in deep ocean environments depend on primary production in the overlying water column. For benthic communities, organic matter must reach the bottom by sinking through a deep water column, a process that further reduces its energy content. Thus, in the open ocean, high rates of secondary production, such as fish yields, are associated with areas in which physical-chemical conditions permit and sustain high rates of primary production over long periods of time, as is found in upwelling regions.

Regardless of specific marine environment, microbial processes provide fundamental links in marine food webs that directly or indirectly govern flows of organic matter and nutrients that in turn control ecosystem productivity and stability.

Marine Habitat

A habitat is the area where an organism or group of organisms live and breed. One habitat will be distinct from another due to its particular environmental conditions. However, habitats are not discrete, and organisms may interact with different habitats within an ecosystem. For example, organisms may move between habitats on a seasonal basis when looking for food or during different parts of their life cycle. Habitats range in size, and their characteristics are determined by a large number of variables. In the marine environment, these variables include light, temperature, substrate, wave action and oxygen availability.

The particular combination of variables results in a habitat suitable for particular types of organisms. For example, the stalk-eyed mud crab is adapted to live in mud flats of an estuary, and cockles are adapted to the fluctuating salinity of the sub tidal zone.

Beaches

A beach is a platform along the shoreline of the sea. A beach, particularly a surf beach, is an unstable habitat. Plants and animals are exposed to constantly changing and often harsh conditions. One of the main obstacles that surf beach-dwelling organisms face is the lack of stable ground. They need to swim or burrow or they will be swept away. Food is limited under these conditions. Organisms eat algae particles growing in the top few centimetres of the substrate or depend on food brought in by the waves.

Estuaries

An estuary is a partially enclosed body of water where freshwater (often from a river mouth) mixes with saltwater from the sea. New Zealand has approximately 300 estuaries, made up of a diverse range of sub tidal and intertidal habitats that vary from vegetated habitats, like sea grass beds, to mud flats, shellfish beds and tidal channels.

Estuaries are typically nutrient-rich, productive areas because of their proximity to land. The differences in ecology between habitats within an estuary are defined by the physical conditions. The dominant physical condition in an estuary is salinity. Other important factors include topography, tidal flow, the amount of freshwater and sediment entering the estuary, turbidity, and wave action and currents.

Each estuarine habitat is home to a diverse range of species. For example, mudflats are full of life – crabs like to burrow through the mud feeding on the microorganisms they come across, many shellfish are partially buried by the mud and help to filter the water by feeding on particulate matter such as tiny planktonic organisms, and bacteria in the mud break down detritus that then provides nutrients for plants that form the base of the food web.

Fiords

A fiord is a steep-sided valley that has been carved out by glaciers and then flooded by the sea. The freshwater input comes from the surrounding landscapes, via waterfalls and run-off from the mountains and native forests. Fiordland National Park in the south-west of the South Island receives over 7.5 metres of rain every year. Large amounts of this freshwater make their way into the fiords, forming a distinct freshwater layer on top of the more dense seawater.

Fiordland has 14 fiords of varying depths. Where the fiord meets the ocean, the depth is typically about 100 metres. Further into the fiord, depths can reach up to 440 metres, but almost all the marine organisms live within the first 40 metres below the surface. The unique climate, vegetation and topography in this part of the country have created a number of specialised underwater habitats within the fiords. The 3 main habitats are defined by depth – the upper zone, the lower zone and deep basins.

The upper zone includes the surface down to 15 metres. Species that can tolerate the freshwater layer thrive in the top 5 metres – these include green seaweed, mussels, barnacles, shrimps and sea stars. Below 5 metres, the number of species increases, and tubeworms, sponges, soft corals, sea squirts, molluscs, sea stars, urchins, sea snails and sea slugs live in great numbers on the steep rock walls.

The lower zone is 15–40 metres below the surface. The majority of the species live on the rock walls and include large sponges, sea squirts, corals, hydrocorals and lampshells. One unique feature of this habitat is the large colonies of black coral that grow here due to the unusual conditions created by the large freshwater layer and low light levels. In other marine environments, black coral grows only at depths below 45 metres.

Below the lower zone at depths of greater than 40 metres, the habitat is suitable for a few species of heart urchins and tubeworms. Below 200 metres, the dark, muddy habitats resemble those in the open ocean at much greater depths and are inhabited by shellfish, heart urchins and crabs.

Continental Shelf

Along with all other major landmasses, New Zealand is surrounded by a gently sloping continental shelf. The continental shelf is an underwater extension of the land and is relatively shallow compared to other areas of the ocean that may be thousands of metres deep. Throughout the world, continental shelves form about 7–8% of the total ocean area, but compared to the rest of the ocean, they are extremely biologically active, well understood and commercially exploited for seafood and minerals.

Black coral: Unusual conditions in Fiordland result in an abundance of black coral in the 15–40 metre zone. In other marine environments, black coral grows only at depths below 45 metres

Continental shelves vary widely in width and depth. The continental shelf around New Zealand ranges from a couple of hundred metres to 100 kilometres in width. The greatest depth is usually 100–160 metres, at which point, it drops sharply at the shelf break and becomes the much steeper continental slope.

The continental shelf provides a number of important habitats for New Zealand's marine species. In the benthic zone, the major influences on habitat type include sediment type (sandy, silty or muddy) and topography (flat, small hills or small valleys).

The numerous habitats on the shelf have been divided into 3 major benthic areas:

- The shallow water zone includes depths of 14–25 metres and is characterised by well-sorted sands.
- The mid-shelf habitat is generally covered in gravel and silt, and reaches to depths of about 90 metres.
- The outer edge of the shelf is from about 90–150 metres deep and typically has a sandy bottom.

For example, off the Otago coastline, species diversity varies across the continental shelf. The mid-shelf habitats have the highest diversity and are dominated by bryozoans and sponges that prefer the rock and gravel seabeds found typically on the mid-shelf.

References

- Marine-biology, science: britannica.com, Retrieved 2 April, 2019
- Marine-microbiology-4604: seattlepi.com, Retrieved 10 January, 2019
- Marine-life-definition-and-examples: thoughtco.com, Retrieved 14 June, 2019
- Marine-ecology, conservation: marinebio.org, Retrieved 17 February, 2019
- Marine-ecology: thefreedictionary.com, Retrieved 7 May, 2019
- 145-marine-habitats: sciencelearn.org.nz, Retrieved 20 March, 2019

2

Marine Fishes, Reptiles and Birds

There are a number of different vertebrate which are studied within marine biology such as fishes, reptiles and birds. Some of the marine reptiles studied within this area of study are sea turtle, sea snake, marine iguana and crocodile. The chapter closely examines these marine animals to provide an extensive understanding of the subject.

Marine Fishes

Marine Fish is a type of fish which is mainly available in marine water and an industry where the marine fishes are devoted for catching, processing, or selling of fish or other aquatic animals are called marine aquarium.

Atlantic Bluefin Tuna

Like many open ocean bony fishes, Atlantic bluefin tuna start out as extremely tiny larvae, no more than a few millimeters long and weighing only a few hundredths of a gram. Within three to five years, individuals reach lengths of three feet (1 m) and are sexually mature. Because Atlantic bluefin tunas undergo such an amazing transformation in size (from being nearly microscopic to being one of the largest open ocean predators), they eat a wide variety of prey throughout their lifetimes. At a young age, they eat tiny zooplankton, and their prey increases in size as they do. As adults, they eat fairly large bony fishes and invertebrates. Similarly, Atlantic bluefin tuna are eaten by a wide variety of predators. When they are newly hatched, they are eaten by other fishes that specialize on eating plankton. At that life stage, their numbers are reduced dramatically. Those that survive face a steady increase in the size of their predators. Adult Atlantic Bluefin is not eaten by anything other than the very largest billfishes, toothed whales, and some open ocean shark species.

Bluefin Tuna are known to be highly migratory, with individuals making long migrations every year. These migrations correspond with their spawning behavior and with their food needs. This species reproduces via broadcast spawning, where several females and several males release millions of eggs and sperm into the water column at the same time. This method increases the likelihood that the eggs will be fertilized and decreases the chances that they will be eaten by egg predators. There are at least two known populations of Atlantic bluefin tuna, one that reproduces in the Gulf of Mexico and one that reproduces in the Mediterranean Sea. Some researchers believe that the Mediterranean population actually represents two populations (one in the west and one in the east plus the Black Sea). This further division has some implications for fisheries management, because the fish in the eastern Mediterranean are generally considered more of a conservation concern than those in the western Mediterranean.

Though almost all fishes are cold blooded, Atlantic Bluefin have a specialized blood vessel structure – called a counter current exchanger – that allows them to maintain a body temperature that is higher than the surrounding water. This adaptation provides them with a major advantage when hunting in cold water, by allowing them to move more quickly and intelligently. The Atlantic bluefin tuna is one of the fastest swimmers in the ocean. Like some shark species, Atlantic Bluefin must constantly swim. In order to obtain oxygen from the water, fishes pass water over their gills. The tunas lack the ability to do so while stopped, so they must continuously swim forward with their mouths open to keep their blood oxygenated.

The Atlantic bluefin tuna is a highly prized food fish and is fished heavily throughout its range. Its value in the high-grade sushi market in Japan and elsewhere is high enough to cause fishers to target this species with almost unmatched effort. Scientists believe the species to be endangered (highly vulnerable to extinction), but fisheries managers continue to allow commercial fishing to target it. While the Atlantic Bluefin was formerly common in the Black Sea and off the coast of Brazil, it has not been observed in significant numbers in either of those places for several decades. As a result of these and other losses, the Atlantic bluefin tuna has experienced the largest range contraction of any open ocean species. Early indications imply that recent strong management regulations may be allowing populations to rebound, but continued conservation and management efforts are required to ensure that more subpopulations are not lost.

Atlantic Wolffish

Atlantic wolffish are voracious predators, and the large head, powerful jaws, and large canine teeth are all used to hunt and eat hard-bodied or spiny invertebrates, such as sea urchins, crabs, large marine snails, etc. They reach lengths of up to five feet (1.5 m).

Atlantic wolffish is usually solitary but form pairs during the breeding season. They mate via internal fertilization, and females spawn relatively large eggs, which both sexes guard until they hatch. This level of parental care is rare among fishes. Individuals have even been known to mate and nest in public aquariums, guarding their eggs for several months.

These fish are restricted to the cold waters of the North Atlantic Ocean, and their blood contains several natural compounds that prevent it from freezing. Atlantic wolffish are good table fare, and throughout its range, it is fished by both recreational fishers and occasional commercial net fishers. It is slow to mature and likely naturally rare, and over the last several decades, scientists have observed steep declines in its numbers. In the United States it is considered a species of concern. As many ecologists consider the Atlantic wolffish to be a keystone species in north Atlantic Ocean food webs, regulating populations of sea urchins and other invertebrates, it is important to consider conservation measures for this rocky reef predator.

Blue Marlin

This species is known for its long bill that grows from the front of its head. Blue marlin use this bill to stun their prey by slashing their heads in a side-to-side motion, knocking potential prey unconscious, and making it easier to catch. The blue marlin's bill differs from the swordfish's and the sailfish's by being round and pointed, rather than flat and blunt. Like many open ocean bony fishes, blue marlin start out as extremely tiny larvae, no more than a few millimeters long and weighing only a few hundredths of a gram. Soon after hatching, they already have a visible bill. Blue marlin grow rapidly, and in the course of their lives they may increase their body weight by at least one million times.

Because blue marlin undergo such an amazing transformation in size (from being nearly microscopic to being one of the largest open ocean predators), they eat a wide variety of prey, throughout their lifetimes. At a young age, they eat tiny zooplankton, and their prey increases in size as they do. As adults, they eat fairly large bony fishes and squids. Similarly, blue marlin is eaten by a wide variety of predators. When they are newly hatched, they are eaten by other fishes that specialize on eating plankton. The size of their predators increases as they grow, and adult blue marlin are not eaten by anything other than large open ocean shark species.

The blue marlin is a highly migratory species, with individuals migrating across entire ocean basins and even between oceans. One individual, for example, was tagged off the east coast of the United States (western Atlantic Ocean) and recaptured in the Indian Ocean. Like many migratory species, the females are larger than males. In fact, adult female blue marlin are as much as four times the size of the largest males. All of the largest individuals are female. Blue marlin reproduce through external (rather than internal) fertilization, where the female releases her eggs into the water column while the male releases his sperm. Once the sperm find the eggs, fertilization occurs. While spawning, a single female may release several million eggs, increasing the likelihood that some will be fertilized and that at least one hatched larva will grow to reach adulthood. Though almost all fishes are cold blooded, blue marlin and the other billfishes have a specialized blood vessel structure – called a counter current exchanger – that allows them to warm their brains and eyes. This adaptation provides them with a major advantage when hunting, by allowing them to think more quickly and see more clearly.

The blue marlin is a targeted fishery species in only a limited number of areas but is captured as accidental bycatch in fisheries targeting other species throughout its range. In some areas, this accidental capture is threatening blue marlin populations, and scientists consider the species vulnerable to extinction. Generally speaking, blue marlin populations are decreasing in numbers. Though it is not targeted commercially throughout much of its range, the blue marlin is one of the most highly sought after big game fishes in the world. Sport fishing enthusiasts spend tens of thousands of dollars to catch large blue marlin, and they are typically released alive (though recent data suggests that released individuals may frequently die after the fact). If numbers continue to decline, this highly valuable sport fishery may be in danger of collapse.

Common Fangtooth

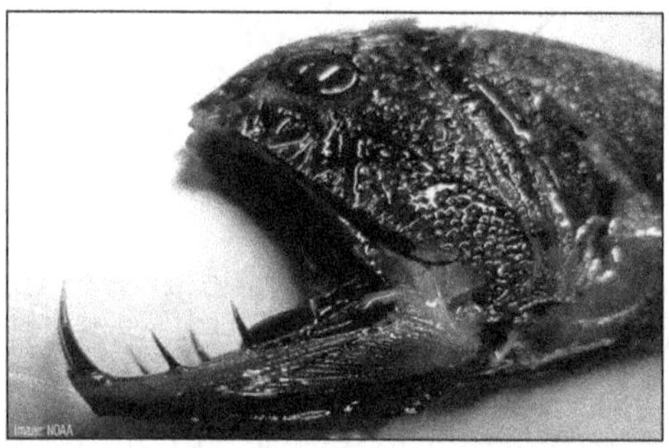

Though they spend most of their time in the deep, common fangtooths are known to migrate toward the surface, at night, following their preferred prey of crustaceans and other fishes. Common fangtooths are more active than many other deep-sea fishes and seek out food, rather than being purely ambush (sit and wait) predators. Their huge mouths and very long teeth ensure that they are able to attack prey that they find in the very large, food-poor deep sea.

Common fangtooths are dark colored – either solid brown or black – and unlike most deep-sea fishes, they do not have light producing organs or cells for communication or to attract prey. Instead, they rely heavily on their sense of smell and benefit from even the slightest bit of sunlight that may make it down to the depths. Though the light is not enough to truly see, it may be enough that potential prey species cast a shadow as they pass overhead. In those circumstances, this species seems to have a bit of a "bite first, ask questions later" attitude.

As they live in an extreme environment where scientists have difficulty studying behavior, little is known about the mating preferences of this species. They are known, however, to reproduce via external fertilization, where females release their eggs and males release their sperm into the open water at the same time.

The common fangtooth, along with the only one other species of fangtooth, is more closely related to shallow-water squirrelfishes than to other deep-sea fishes. Though they look menacing and are voracious predators, common fangtooths are harmless to humans. They live in the deep sea and only grow to about 7 inches (18 cm) long.

Guineafowl Puffer

The guineafowl puffer, like all puffers, has powerful jaws and beak-like teeth. The teeth of the upper jaw are fused into two tooth plates, as are the teeth in the lower jaw. In fact, the family name for puffers (Tetraodontidae) means "four teeth," reflecting the two tooth plates that these fishes have in each jaw. Guineafowl puffers are foraging predators that eat a variety of sessile invertebrates. Their preferred prey is branching coral (including lobe coral), but they also occasionally eat sponges, soft corals, and other reef organisms.

The guineafowl puffer is a poor swimmer and can easily be caught by predators. However, when threatened, this species has the incredible ability to swallow water and inflate itself to several times its normal size. Though the process takes a few seconds, it is quite effective at preventing predation. It is very difficult for a predator to swallow an inflated individual. When a guineafowl puffer no longer feels threatened, it uses muscles along its abdomen to push the water back out of its mouth. Even if successfully eaten (before it can inflate), this species can be poisonous to potential predators, as it stores toxins created by symbiotic bacteria that live inside its body. Therefore, this species is rarely eaten.

The guineafowl puffer is typically fairly uncommon but can reach large numbers at some locations. It is generally not eaten (partly as a result of its toxic nature) but is a valuable species in the live trade, where it is captured for display in public and private aquaria. Its numbers may have been reduced in some locations, but populations are generally stable, and the guineafowl puffer is considered to be a species of least conservation concern.

John Dory

The John Dory is an active predator and eats a variety of schooling fishes and invertebrates. It lives in a wide depth range, from 15 feet (5 m) to 1200 feet (360 m) and usually stays near the seafloor, over both soft and hard bottoms. John Dories are medium-sized predators in the systems in which they live. They reach lengths of approximately two feet (65 cm) and weights of a few pounds. Larger bony fishes and sharks are known to feed on John Dories.

This species is typically somewhat solitary but comes together in groups to reproduce. It spawns by a behavior known as substrate scattering, where females scatter their eggs, haphazardly, on the seafloor, and males release their sperm in the same location to fertilize them externally. Individuals of this species do not construct nests and do not provide any care of the eggs. Unlike the eggs of species that reproduce via broadcast spawning, the eggs of the John Dory and other substrate scattering fishes do not float. Only after the eggs hatch do the larvae swim up into the water column to live a planktonic lifestyle.

The John Dory is generally considered an excellent food fish and is targeted by commercial fisheries in some areas. Its conservation status is not currently known, but it is not fished as heavily as some closely related species. However, this species is targeted by bottom trawling, a method known to be highly destructive to benthic habitats. It is therefore important to carefully manage the fishing activities for the John Dory, in order to ensure that this species' habitat is not irreversibly damaged.

Sailfish

Sailfish are a type of billfish (like the blue marlin or swordfish) that are known not only for their pointed bills, but also their extraordinary dorsal fins that can be taller than the length of their bodies. Sailfish start out as tiny larvae, no more than a few millimeters in length, but grow rapidly during their first year. They can grow from 0.125 inches (0.3 cm) when born to 10 feet (3 m) long from bill to tail and weigh upwards of 220 pounds (100 kg) in adulthood. Like other billfish, female sailfish can be larger than males so they can carry as many eggs as possible and have a greater chance of successful reproduction in the open ocean.

Sailfish eat a wide variety of prey throughout their lifetimes. At a young age, they eat tiny zooplankton, and their prey increases in size as they do. As adults, they eat fairly large bony fishes, crustaceans and squid. Sailfish also work together, using their dorsal fins to create a barrier around their prey, in order to feed on smaller schooling fish, such as sardines and anchovies.

Sailfish are eaten by a wide variety of predators. When newly hatched, sailfish are preyed on by other fishes that specialize on eating plankton. The size of their predators increases as they grow, and adult sailfish are not eaten by anything other than larger predatory fish like open ocean shark species, orcas and dolphinfish.

During spawning, a female will attract a male partner by extending her dorsal fin above the surface of the water. Male-female pairs form and reproduce through external fertilization, where the female releases her eggs into the water column while the male releases his sperm. While spawning, a single female may release several million eggs to increase the likelihood that some will be fertilized.

Sailfish are popular in recreational fishing (catch and release) but have little value in the commercial fishing industry and cannot be fished commercially in the Atlantic Ocean. Sport fisheries account for the sailfish's highest catch rates, particularly in the Eastern Pacific Ocean where the species can be found near Central American coasts. Sailfish are also caught as bycatch by driftnets, harpoons and commercial, long-line tuna fisheries, where bycatch numbers are not accurately reported.

Scorpionfish

Scorpionfish are bottom-dwelling fish that have also been called rockfish or stonefish because of their tendency to live among rocks near the seafloor. There are more than 200 known species of scorpionfish in the ocean. Members of this fish family are commonly found in the Indian and South Pacific Oceans where water temperatures are temperate and coral reefs are plentiful. Coral reefs provide the perfect space for a scorpionfish to hide and hunt for prey and also avoid any potential predators brave enough to take a bite. But, the body of a scorpionfish is just as important as its habitat when it comes to remaining unseen. Scorpionfish are covered in feathery fins or skin flaps that help with camouflage against surrounding coral. Some scorpionfish are dull in color–mottled brown or yellow– while other species are bright red or orange, making them virtually invisible when hidden among either rocks or reefs. Scorpionfish are also equipped with spines containing dangerous venom. When the spines pierce a predator, the venom is injected immediately at the point of contact. A sting from one of these spines can be potentially fatal to other animals and extremely painful to humans.

Scorpionfish are exceptional sit-and-wait predators. Scorpionfish only hunt at night and spend daylight hours resting in crevices. They will remain in the shadows of rocks or reefs before pouncing on unsuspecting prey swimming by. Their diet consists of small fish, crustaceans and snails that also live in coral reefs. A scorpionfish's mouth is wide in size, which allows the fish to quickly suck and swallow prey whole in one bite. Scorpionfish can also use their venom to stun prey before eating them. Predators of scorpionfish remain few, but sharks, rays and large snappers have been known to hunt the fish.

Most scorpionfish species are less than two feet in size and have a lifespan of 15 years. A female scorpionfish may release upwards of 15,000 eggs into the water for fertilization by the males. Fertilized eggs float to the surface where egg-eating predators are less likely to get to them. After just two days the eggs will hatch and baby scorpionfish will remain near the surface until they are big enough to swim down to the reefs where the adults swim. Scorpionfish face few threats in our ocean—they are not commercially fished because of their venom, but some species may be caught as incidental bycatch when lines and hooks are set near reefs. Their coral reef habitat may also be in jeopardy due to warming oceans and global warming.

Marine Reptiles

The marine reptiles come from ancient terrestrial forms that eventually colonized the sea. The number of true marine species represents only 1% of all the reptile species that exist today. The true marine species are sea turtles, Marine Iguana and sea snakes. Some other species of turtles, lizards, snakes and crocodiles have estuarine habits, but always need freshwater to some extent.

The fossil record tells us that the marine environment became colonized several times by different groups of reptiles. Some of the families involved only contain marine species. During the Mesozoic they achieved a great diversity of forms.

The marine environment has a high content of salt. For that reason the marine reptiles have developed adaptive mechanisms to limit salt intake and to facilitate its excretion. Other physiological adaptations are related to diving and to thermoregulation. Also some anatomical adaptations have been needed to move efficiently in the water.

Physiological Adaptations to Sea Life

The marine reptiles had to adapt to an aquatic environment which is full of salt and had to develop a breathing capacity adapted to diving. They also had to solve the locomotory problem in an environment which has a higher density and viscosity than air. Nevertheless, some of their physiological adaptations that already had developed for their prior life on land helped them during their new colonizing process. These helpful reptile adaptations were the ectothermy, the capacity for anaerobic metabolism, tolerance of acid-base disturbance, ventilation with apneic periods, and a circulation system with intra-ventricular shunting.

Salt and Water Balance

Seawater has a complex chemical composition, with many dissolved salts, of which chlorine and sodium ions are the most abundant. But there are also important quantities of magnesium, potassium, strontium, bromine, fluorine, sulfate and carbonate ions. And there are more ions in few quantities. Animals cannot accumulate these ions in their bodies, because they will become toxic if they reach certain levels. For that reason, marine reptiles need to maintain water balance to survive. They attain this by reducing losses of body water and by facilitating salt excretion. Body water is normally lost while breathing, through the skin and through excretion (salt glands, urine and feces), and salt comes into the body during feeding and through diffusion.

Although the skin is a water-permeable tissue, it plays an important role in osmotic regulation and in ionic balance. In marine and estuarine species this permeability is smaller than in freshwater species, to avoid dehydration process due to the salt rich environment. In addition the skin permeability is more resistant to the passage of water outward from the body. Another important feature of reptiles' skin is the ionic impermeability, which avoids salt entry, especially sodium.

On the other hand, the salt concentration in the marine reptile's body is less than that in the surrounding seawater. In contrast to other marine species, reptiles cannot tolerate high salt levels in their bodies. This would kill them. However, the plasma's osmotic concentration of electrolytes and nitrogenous compounds is higher in marine reptiles than in freshwater ones. This is because of the high levels of sodium, chloride and urea. They accumulate these ions and the urea in the extracellular fluid. When marine or estuarine reptiles are dehydrated, it is commonly due to sodium accumulation (hypernatremia) and not to potassium accumulation (hyperkalemia), because high concentrations of potassium are toxic to them. Unusual high concentrations of intracellular potassium will damage the activity of the enzyme systems and synthesis of DNA, RNA and protein.

A good system of eliminating salt excess while avoiding water loss is excreting very concentrated urine. Mammals accomplish this by producing urine which is more concentrated in salts than their body fluids, using a counter-current mechanism in the kidney. But the kidney of reptiles is not able to produce urine more concentrated than the blood. Marine reptiles regulate salt content by the means of specific salt glands. Sea snakes, sea turtles, crocodiles, Marine Iguana and some lizards have salt glands, but they are all a result of convergent evolution. They are not a homologous salt gland. The Marine Iguana *Amblyrhynchus cristatus* and some lizards use a nasal gland; the sea turtles and Diamondback Terrapin use a lachrymal gland; the true sea snakes (Hydrophiidae), the sea kraits (Laticaudidae) and the *Acrochordus* use a sublingual gland; the Cerberus use the premaxillary gland; and the crocodiles use a lingual gland. These glands excrete mainly sodium and chloride ions, but also potassium and bicarbonate ions. The Marine Iguana excretes great quantities of potassium, which is an abundant ion in their diet but which is toxic in big concentrations.

As a general rule, marine reptiles do not drink sea water. They obtain the necessary water out of their food. Estuarine species and some sea snakes, however, do drink when they have freshwater available.

Finally, there are some species that use estuarine environments, but are not marine species, like some turtles, lizards, crocodiles and snakes. They select a favorable microhabitat where salt content is smaller than in the sea. When salinity is low, they stay in the estuarine habitat. But when the salinity increases they migrate to the rivers and freshwater areas. An example of migrating behavior to control salinity is the case of the Painted Terrapin *Callagur borneoensis*. This species does not have a salt gland, but the adults can tolerate seawater, so they nest in sea beaches. However, the hatchlings migrate immediately to the rivers after being born.

Respiration and Diving Adaptations

Respiration in reptiles has characteristic apnea periods, whose duration vary from one species to the other and depends on the circumstances. In aquatic reptiles the apneic periods are longer than in terrestrial species. These apnea periods allow them to stay a long time underwater with no effort. Because reptiles are ectothermic organisms and produce little metabolic heat, they need less oxygen than a mammal or a bird of the same size—another point that allows them to prolong their sojourns underwater.

Their breathing cycles present a characteristic series of rapid breaths, followed up by an apneic period. This will last until a new series of rapid breaths come. Sea turtles and sea snakes take only one breath each time they surface. Sea kraits show different breathing rythms when they are in the sea than when they are on land. In the water they have long apneic periods followed by several rapid breaths, but on land they have frequent and regular breaths.

Sea snakes can breathe through the skin, in addition to their pulmonary respiration. Through the skin they obtain around a fifth of the oxygen they need and almost all the produced carbon dioxide is eliminated through the skin. This latter feature is very important during the immersions, because it avoids its over-concentration in blood. Also nitrogen elimination through the skin can reduce the risk of bubble formation in long dives that would produce decompression sickness.

Marine reptiles can control their buoyancy and maintain a position in the water column without spending extra energy in active swimming. They can stay on the bottom (negative buoyancy), at certain depth (neutral buoyancy) or on surface (positive buoyancy) just by getting air in or out of their lungs, just like the buoyancy control device of a scuba diver.

For example, the Yellow-bellied Sea Snake Pelamis platurus can control its buoyancy with its elongated lung. It can keep positive buoyancy and remain at the surface, or stay neutral at four meters depth. The Marine Iguana reaches the neutral buoyancy when it is at four to five meters depth.

Marine reptiles generally do their immersions under conditions of aerobic metabolism. But when the dives last very long or are very repetitive in a short period of time, the metabolism becomes anaerobic. For example, the Marine Iguana does its short feeding immersions using aerobic metabolism, but during the longer escape dives, the metabolism can become anaerobic. When sea turtles hibernate, they stay completely submerged or buried in the mud for weeks. In this case the anaerobic metabolism plays an important role.

Thermoregulation

Reptiles are ectothermic animals, but their life cycles take place in a range of optimal body temperatures. When under or over this range the reptiles will cease their activities. And if the lower or upper critical threshold is surpassed, they will die. As with all other reptiles, they have some control of their body temperature—it is not always the same as the temperature of the surrounding environment. They can increase or decrease their body temperature through certain behaviour, like basking on the surface or like selecting microhabitats at lower or higher temperatures.

But even when reptiles have low rates of metabolic heat production, they can also produce some internal body heat and keep it, thanks to a layer of insulating fat. For example, the Leatherback Turtle Dermochelys coriacea can have a body temperature 18 °C above the sea temperature. This can be done thanks to the large size that they attain. They are the largest present marine reptiles, reaching 2.4 meters and weighting 900 kg.

Their favorable surface-volume relation helps them to reduce body heat losses, as does their thick layer of fat under the carapace and their counter-current circulatory system at the fins. This capacity allows them to remain active in waters at just 6 to 8 °C. The Green Turtle Chelonia mydas, can maintain a temperature 8 °C over that of the surrounding sea water. This thermic regulation capacity is less developed in smaller sea turtle species.

Sea snakes like the Yellow-bellied Sea Snake can have a body temperature 2.4 °C above the sea temperature. Anyway, since all the species from this group live in tropical or subtropical waters, their body temperature is very similar to the sea temperature. The sea kraits can get out of the water to bask in the sun and regulate their temperature. The Yellow-lipped Sea Krait Laticauda colubrina maintains its body temperature between 29 to 31 °C when it is on land, even when the surrounding temperature does change. This thermic regulation is due to its behaviour. It selects the cooler places during the hottest hours, and exposes itself to the sun to increase body temperature, when the temperature drops down.

The Marine Iguana controls its body temperature by basking in the sun on the shore and reaches up to 35 to 37 °C. The sea in the Galápagos islands is cold. So, when the iguana does a foraging trip to the sea, its body temperature drops, to as low as 25 °C. It has two strategies to control its body temperature. The smaller animals do short trips, returning quickly back to the shore. By doing so, they keep their body temperature more or less stable. The bigger animals do longer trips and can stay in the sea until their body heat drops down to 25 °C.

Sea Turtle

Sea turtle (or seaturtle) is the common name for any of the large marine turtles comprising the superfamily Chelonioidea, characterized by forelimbs in the form of large flippers or paddles. There are two extant families, Cheloniidae and Dermochelyidae. Members of the family Cheloniidae are characterized by a lightweight, low shell covered with scutes, while the sole extant species in Dermochelyidae, the leatherback sea turtle (Dermochelys coriacea), are characterized by a teardrop shaped carapace lacking epidermal scutes and covered with a leathery skin. Members of both families have a large head that cannot be retracted into the shells. There are seven living species, arranged into six genera. Three extinct genera also are recognized. Sea turtles are found worldwide in all the world's oceans except perhaps the Arctic Ocean.

Sea turtles provide important functions for the ecosystem. They are a significant part of food chains, both in the ocean and on land where they lay many eggs, with both the eggs and the hatchlings particularly susceptible to predation. Sea turtles also recycle nutrients to the beach and dune ecosystems, which helps to support vegetation and prevent erosion. And, while generally carnivorous, those that graze sea grass help to keep this vegetation healthy.

Sea turtles have historically provided an important function for people as food and with the shells used for decorative purposes. However, today the main function is one of adding to the wonder of nature and ecotourism trips. Various anthropogenic factors have lead to serious declines in sea turtle populations such that all are considered threatened, with three species classified as critically endangered and two as endangered. As a result, various protective mechanisms have been put in place for conservation of these great reptiles, but poaching, accidental death during fishing, and habitat destruction, among other factors, continue to threaten their populations.

Sea turtles, along with other turtles and tortoises, are part of the Order Testudines. Although sea turtles have been around for tens of millions of years since the Mesozoic, the body plan of sea turtles has remained relatively constant. Sea turtles possess dorsoventrally-flattened bodies with two hind legs and highly-evolved paddle-like front arms. Extant sea turtles are placed into two families within the superfamily Chelonioidea.

Cheloniidae includes six species in five genera: flatback *(Natator depressus)*, green sea turtle *(Chelonia mydas)*, hawksbill *(Eretmochelys imbricata)*, Kemp's ridley *((Lepidochelys kempii))*, olive ridley *(Lepidochelys olivacea)*, and loggerhead *(Caretta caretta)*. The East Pacific subpopulation of the green turtle was previously classified as a separate species, the black turtle, but DNA evidence indicates that it is not sufficiently distinct from the green turtle. These species are all characterized by a streamlined shell that is low and covered with scutes (external plates derived from the epidermis), paddle-like forelimbs, a large head that cannot be retracted into the shell, and a skull with a solid, bony roof. Different species are distinguished by varying anatomical aspects: for instance, the prefrontal scales on the head, the number of and shape of scutes on the carapace, and the type of inframarginal scutes on the plastron. Species generally range from two to four feet in length (0.5 to 1 meters) and proportionally narrower. The olive ridley is the smallest sea turtle, with an average weight just over 100 lb (up to 50 kilograms) and a carapace length reaching only 30 inches (70 centimeters).

Dermochelyidae includes one extant species, the leatherback sea turtle *(Dermochelys coriacea)*. The leatherback is the only sea turtle that does not have a hard shell, instead carrying a mosaic of hundreds of bony plates just beneath its leathery skin. It also is characterized by a smooth, streamlined carapace that is teardrop shaped, seven longitudinal ridges, no epidermal scutes, no scales on the head, and a prominent tooth-like cusp on both sides of the upper jaw (Iverson 2004b). The paddle-like forearms lack claws (Iverson 2004b). The leatherback is the largest of the sea turtles, measuring six or seven feet (2 meters) in length at maturity, and three to five feet (1 to 1.5 m) in width, weighing up to 1300 pounds (650 kg).

Green turtle breaks the surface to breathe

Sea turtles spend almost all their lives submerged but must breathe air for the oxygen needed to meet the demands of vigorous activity. With a single explosive exhalation and rapid inhalation, sea turtles can quickly replace the air in their lungs. The lungs are adapted to permit a rapid exchange of oxygen and to prevent gasses from being trapped during deep dives. The blood of sea turtles can deliver oxygen efficiently to body tissues even at the pressures encountered during diving. During routine activity, green and loggerhead turtles dive for about 4 to 5 minutes and surface to breathe for 1 to 3 seconds.

Sea turtles possess a salt excretory gland at the corner of the eye, in the nostrils, or in the tongue, depending upon the species; chelonian salt glands are oriented in the corner of the eyes in leatherback turtles. Due to the iso-osmotic makeup of jellyfish and other gelatinous prey upon which sea turtles subsist, sea turtle diets are high in salt concentrations and chelonian salt gland excretions are almost entirely composed of sodium chloride at approximately 1500-1800 mosmoll-1.

Turtles can rest or sleep underwater for several hours at a time but submergence time is much shorter while diving for food or to escape predators. Breath-holding ability is affected by activity and stress, which is why turtles drown in shrimp trawls and other fishing gear within a relatively short time (MarineBio).

Species

Immature Hawaiian Green turtle in shallow waters

Seven distinct extant species of sea turtles are recognized; they constitute a single radiation that was distinct from all other turtles at least 110 million years ago. During that radiation, sea turtles split into two main subgroups, which still exist today: The unique family *Dermochelyidae,* which consists of a single species, the leatherback; and the six species of hard-shelled sea turtle, in the family *Cheloniidae.*

- Family Cheloniidae:
    - Chelonia mydas Green sea turtle
    - Eretmochelys imbricata Hawksbill turtle
    - Natator depressus Flatback turtle
    - Caretta caretta Loggerhead sea turtle
    - Lepidochelys kempii Kemp's Ridley
    - Lepidochelys olivacea Olive Ridley or Pacific Ridley.
- Family Dermochelyidae:
    - Dermochelys coriacea Leatherback sea turtle.

Distribution

The superfamily Chelonioidea has a worldwide distribution excluding near the poles. Most sea turtles are found in tropical and semi-tropical regions, but also are found in temperate regions. The leatherback sea turtle, which has a worldwide distribution, is the only turtle, or reptile, to regularly travel to and survive Arctic or sub-Arctic habitats; it has been sighted regularly off Newfoundland, Canada and observed off the shores of Greenland as far north as 69 °N. It also is one of two turtles to regularly nest in the Mediterranean, along with the green turtle, although the hawksbill and Kemp's ridley turtle occur there occasionally. The leatherback, which may have the greatest distribution of any reptile, also is found in cold temperate waters around Newfoundland, Iceland, Alaska, Norway, New Zealand, and Cape of Good Hope. The Flatback turtle is found solely on the northern coast of Australia.

Behavior, Life History, Diet and Predation

Green turtle swims above corals at Hawaii

A feeding sea turtle,Chelonia mydas

Members of the Cheloniidae family tend to enter temperate seas during the summer, but migrate to warmer waters for the winter, or bury themselves in the mud in shallow coastal areas during the winter. The leatherback sea turtles can feed in both cold and warm environments because they can maintain their body temperature above the water temperature. They generate body heat by muscle activity and it is lost very slowly, their oily skin has excellent insulative properties, and they have a countercurrent circulatory system that helps in heat retention. Leatherback turtles can dive to depths of more than 3,300 feet (1,000 meters) to reach food sources.

Reproduction and Life History

Although sea turtles spend most of their lives in the oceans, they come ashore to nest and lay their eggs. They often make extremely long migrations between their feeding grounds and their nesting grounds. Some migrate up to 190 miles (300 kilimeters) and leatherback sea turtles have been recaptured more than 3,100 miles (5,000 kilometers) from where released, with individuals traveling more than 19 miles (30 kilometers) per day for several weeks. The ridley turtles are especially peculiar because instead of nesting individually like the other species, they come ashore in one mass arrival known as an "arribada" (arrival). With the Kemp's ridley, this occurs during the day. Their numbers used to range in the thousands but due to the effects of extensive egg poaching and hunting in previous years the numbers are now in the hundreds. Most nesting is done at night.

Sea turtles are highly sensitive to the Earth's magnetic field and apparently use it to navigate. The longevity of sea turtles has been speculated at 80 years. The fact that most species return to nest at the locations where they were born seems to indicate an imprint of that location's magnetic features.

After about 30 years of maturing, adult female sea turtles return to the land to nest, usually on the same beach from which they hatched. This can take place every two to four years in maturity. Sea turtles make from four to seven nests per nesting season.

All sea turtles generally employ the same methods when making a nest. A mature nesting female hauls her onto the beach until she finds suitable sand on which to create a nest. Using her hind flippers, the female proceeds to dig a circular hole 40 to 50 centimeters deep. After the hole is dug, the female then starts filling the nest with a clutch of soft-shelled eggs, one by one, until she has deposited around 150 to 200 eggs, depending on the turtle's species. The nest is then re-filled with loose sand by the female, re-sculpting and smoothing the sand over the nest until it is relatively undetectable visually. The whole process takes around thirty minutes to a little over an hour. After the nest is laid, the female then returns to the ocean.

Some of the eggs are unfertilized and the rest contain young turtles. Incubation takes about two months. The length of incubation and the gender of the hatchling depend on the temperature of the sand. Darker sands maintain higher temperatures, decreasing incubation time and increasing the frequency of female hatchlings. When the time comes, these hatchlings tear their way out of their shells with their snout and once they have reached the surface of the sand, they will instinctively head towards the sea. Only a very small proportion of them (usually .01 percent) will be successful, as many predators wait to eat the steady stream of new hatched turtles (since many sea turtles lay eggs en masse, the eggs also hatch en masse).

The hatchlings then proceed into the open ocean, borne on oceanic currents that they often have no control over. While in the Open Ocean, it used to be the case that what happened to sea turtle young during this stage in their lives was unknown. However in 1987, it was discovered that the young of *Chelonia mydas* and *Caretta caretta* spent a great deal of their pelagic lives in floating sargassum beds—thick mats of unanchored seaweed floating in the middle of the ocean. Within these beds, they found ample shelter and food. In the absence of sargassum beds, turtle young feed in the vicinity of upwelling *"fronts"*. Reports in 2007 verified that green turtle *(Chelonia mydas)* hatchlings spend the first three to five years of their lives in pelagic waters. Out in the open ocean, pre-juveniles of this particular species were found to feed on zooplankton and smaller nekton before they are recruited into inshore seagrass meadows as obligate herbivores.

Diet

With the exception of the largely herbivorous green sea turtle, which feeds mainly on sea grasses, sea turtles are primarily carnivorous. Prey includes mollusks, crustaceans, sponges, sea urchins, barnacles, and fish. Leatherbacks prefer oceanic jellyfish, but also consume snails, bivalve mollusks, octupuses, squids, crabs, hydrozoans, tunicates, and small fish, and supplement their diet with algae, kelp, and sea grasses.

Most researchers' theories portray that sea turtles have an acute sense of smell in the water. Their experiments showed that the hatchlings reacted to the scent of shrimp. This sense allows sea turtles to locate food in deep and murky water.

Sea turtles open their mouths a bit and draw in water through the nose, then immediately empty it out again through the mouth. Pulsating movements of the throat are thought to be associated with smelling.

Predators and Protections

Like many other animals in the world, sea turtles have predators. While adult sea turtles have few predators, mainly sharks, the eggs and hatchlings face many predators, including fish, seabirds, raccoons, and lizards. One natural protection is their shell. Other protections include the ability of some species' massive jaws to suddenly snap shut, and to stay underwater for hours on end; these are both instinctual and natural.

Turtles have many senses to aid them in the sea. Sea turtle ears have a single bone in the middle ear that conducts vibrations to the inner ear. Researchers have found that sea turtles respond to low frequency sounds and vibrations. Sea turtles have an extremely good eyesight in water but are shortsighted on land. Under experimental conditions, the loggerhead and green sea turtle hatchlings showed a preference for ultraviolet, blue-green and violet light. Sea turtles are touch-sensitive on the soft parts of their flippers and on their shell.

Ecological Importance

Sea turtles on a beach in Hawaii

In addition to their key role in food chains, sea turtles play key roles in two ecosystems that are critical to them as well as to humans: The oceans and beaches/dunes. If sea turtles were to become extinct, the negative impact on beaches and the oceans would potentially be significant.

In the oceans, for example, sea turtles, especially green sea turtles, are one of the very few creatures (manatees are another) that eat a type of vegetation called sea grass that grows on the sea floor. Sea grass must be kept short to remain healthy, and beds of healthy sea grass are essential breeding and development areas for many species of fish and other marine life. A decline or loss of sea grass beds would mean a loss of the marine species that directly depend on the beds, which would trigger a chain reaction and negatively impact marine and human life. When one part of an ecosystem is destroyed, the other parts will follow.

Beaches and dunes are a fragile ecosystem that does not get many nutrients to support its vegetation, which is needed to help prevent erosion. Sea turtles contribute nutrients to dune vegetation from their eggs. Every year, sea turtles lay countless numbers of eggs in beaches during nesting season. Along one twenty-mile (32 kilometer) stretch of beach in Florida alone, for example, more than 150,000 pounds of eggs are laid each year. Nutrients from hatched eggs as well as from eggs that never hatch and from hatchlings that fail to make it into the ocean are all sources of nutrients for dune vegetation. A decline in the number of sea turtles means fewer eggs laid, less nutrients for the sand dunes and its vegetation, and a higher risk for beach erosion.

Importance to Humans

Moche Sea Turtle

Marine turtles are caught worldwide, despite it being illegal to hunt most of the species in many countries. A great deal of intentional marine turtle harvests worldwide are for the food industry. In many parts of the world, the flesh of sea turtles is considered fine dining. A text dating back to the fifth century B.C.E. describes sea turtles as exotic delicacies in ancient China. Historically, many coastal communities around the world have depended on sea turtles as a source of protein. Several turtles could be harvested at once and kept alive on their backs for months until needed, including on the deck of ships. The skin of the flippers is also prized for use as shoes and assorted leather-goods.

To a much lesser extent, specific species of marine turtles are targeted not for their flesh, but for their shells. Tortoiseshell, a traditional decorative ornamental material used in Japan and China, is derived from the carapace scutes of the hawksbill turtle. The use of marine turtle shells for decorative purposes is by no means limited to the orient. Since ancient times, the shells of sea turtles (primarily the hawksbill) have been used by the ancient Greeks and ancient Romans. Various articles and ornaments used by the elite of these societies, such as combs and brushes, were from processed turtle scutes. The Moche people of ancient Peru worshiped the sea and its animals. They often depicted sea turtles in their art.

Conservation

All species of sea turtles are listed as threatened species according to the IUCN Red List. The leatherback, Kemp's ridley, and hawksbill turtles are listed as critically endangered. The olive ridley and green turtles are considered endangered, and the loggerhead is a threatened species. The flatback's conservation status is unclear due to a lack of research data.

There are a number of anthropogenic factors that have contributed to the decline in sea turtle populations, including hunting, gathering eggs, fishing bycatch and accidental deaths, and habitat destruction. Sea turtles used to be hunted on a large scale in the whaling days for their meat, fat, and shells. Coastal peoples have also been known to gather turtle eggs for consumption.

Legal notice

One of their most significant threats now comes from bycatch and accidental deaths due to various fishing methods. Long-line fishing is cited as one of the causes of sea turtle deaths. Likewise, nets used in shrimp trawling and fishing have been known to cause the accidental deaths of sea turtles. The turtles, as air-breathing reptiles, must surface to breathe. Caught in a fisherman's net, they are unable to go to the surface to breathe and suffocate to death in the net. Some relatively inexpensive changes to fishing techniques, such as slightly larger hooks and traps from which sea turtles can escape, can dramatically cut the mortality rate. Turtle Excluder Devices (TEDS) have reduced sea turtle bycatch in shrimp nets by 97 percent. Another danger comes from marine debris, especially from abandoned fishing nets in which they can become entangled.

Beach development is another area which poses threats to sea turtles. Since sea turtles return to the same beach locations to nest, if these areas are developed they may be left with nowhere to nest, or their nesting locations may be threatened by human activity. Therefore, there has been a movement to protect these areas, in some cases by special police. In some areas, such as the east coast of Florida, after the adult turtles lay their eggs, they are dug up and relocated to special fenced nurseries where they can be protected from beach traffic. (This is not the best thing to do, as many turtle species return to the beach on which they were born.) Hatchlings find their way to the ocean by crawling towards the brightest horizon, but often become disoriented on developed stretches of coastline. Special lighting ordinances may also be enforced to prevent lights from shining on the beach and confusing young hatchlings, causing them to crawl towards the light and away from the water, usually crossing a road. A turtle-safe lighting system uses red light in place of white light as sea turtles cannot see red light.

A green sea turtle at rest

Another major threat to sea turtles is the black market trade in eggs and meat. This is a pervasive problem throughout the world, but especially a concern in the Philippines, India, Indonesia, and throughout the coastal nations of Latin America. Estimates are as high as 35,000 turtles killed a year in Mexico and the same number in Nicaragua. Conservationists in Mexico and the United States have launched "Don't Eat Sea Turtle" campaigns in order to reduce the urban black market trade in sea turtle products. Sea turtles are often consumed during the Catholic holiday, Lent, even though they are reptiles, not fish. Conservation organizations have written letters to the Pope asking that he declare turtle's meat.

Sea turtles can also be affected by Fibropapillomatosis, a disease that has been found among sea turtle populations and causes tumors.

Sea Snake

Besides the sea turtles, the sea snakes are the reptiles that are best adapted to marine environments. The most typical feature of a sea snake is the vertically flattened paddle-like tail, which is not found in any other terrestrial or aquatic snakes. Sea snakes occur in the tropical and subtropical waters of the Indian and Pacific Oceans from the east coast of Africa to the Gulf of Panama. However, two specimens of *Pelamis platurus* reported from Namibia indicate that the species may be extending its range into the Atlantic Ocean. Most species are concentrated in the Indo-Malayan Archipelago, South China Sea, Indonesia and the Australian region. Sea snakes inhabit shallow waters along coasts and around islands, river mouths and up rivers for more than 150 km and they have also been found in lakes in Thailand, Cambodia, the Philippines and Rennell Island in the Solomon archipelago. However, information on precise geographical distribution and abundance for each species is still lacking.

A typical sea snake, Hydrophis belcheri

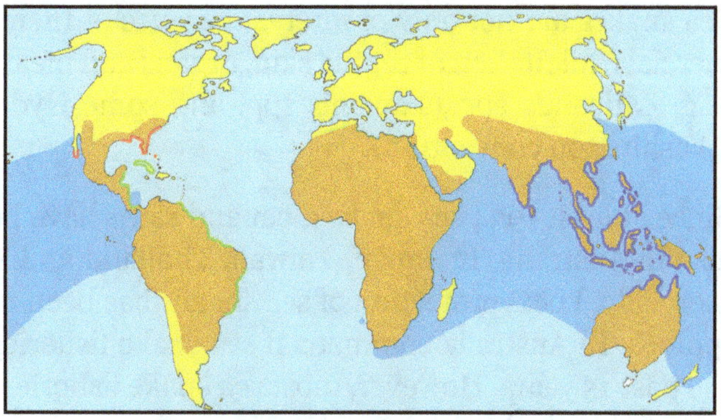

Distribution of marine snakes

Terrestrial distribution represents terrestrial elapids (brown), marine distribution represents sea snakes, i.e. the subfamily Hydrophiinae of the Elapidae (blue). Dark blue: homalopsid snakes along the Asian and Australian coasts. Red: North-American Natricidae, green: neotropical Dipsadidae.

Notably, a few sea snakes are actually freshwater species, such as Hydrophis semperi, H. sibauensis, H. obscurus, and L. crockeri. However, all evidence indicates that these species have radiated into freshwaters independently from saltwater species.

Sea Snake Bite

Sea snake bite is the cause of human fatalities. The typical victim is a fisherman handling gape nets or down nets, sorting fish (and sea snakes) on board a trawling boat or dragging a net by wading in muddy coastal waters or in river-mouths. Some sea snakes are gentle, inoffensive creatures that bite only under provocation, but other species are much more aggressive (e.g. Aipysurus laevis, Astrotia stokesii, Enhydrina schistosa, Hydrophis elegans, H. macdowelli, H. major, H. ornatus and H. ocellatus). As a general rule all sea snakes must be handled with great caution: however, it is worth mentioning that when sea snakes do bite, they do not always inject much or any of their venom, so that only trivial severity of poisoning will be recognizable.

Feeding and Breeding Biology

Most sea snake species feed on fish that are close to the bottom or sedentary; a few prefer fish eggs (Aipysurus eydouxii, and the genus Emydocephalus; some specimens of A. fuscus also contained fish eggs) and at least Aipysurus laevis, Enhydrina schistosa and Lapemis curtus have been found with crustaceans and mollusks in their stomachs. The Laticauda group feed mostly on eels burrowed in sand at the bottom of the sea and in reef crevices. Using sea snakes to capture undescribed eel species has been shown to be very effective because the eels are extremely secretive in their habits giving no chance to collect those using traditional methods.

All sea snakes produce living young (viviparous) except for the genus Laticauda which is egg laying (oviparous). The annual reproductive cycles are synchronous between males and females and they reproduce every second year with a clutch size that increases with the size of the female.

By-catch and Commercial use of sea Snakes

Sea snakes are not only of interest because of their poison, but also in connection with the commercial exploitation of reptile skin, organs and meat. Some species are accessible in great numbers (e.g. Laticauda spp., Lapemis spp. and some Hydrophis spp.) and are not protected by CITES (Washington convention).

Since at least 1934 sea snake meat and skins have been used commercially in the Philippines, but also in Australia. In Japan, Taiwan, Thailand and Vietnam sea snakes have been collected commercially. Local protection of sea snakes has been necessary to stop over-exploitation in the Philippines. In Australia commercial sea snake fisheries and by-catch have been investigated during the past 15 years. However, most sea snake fisheries in the Indian Ocean and in the Pacific are not reported in the literature and are beyond control of the local governments.

Species Distribution and Density

Many of the more than 60 species of sea snakes have a broad distribution in both the Indian Ocean and the Pacific Ocean. Species such as Acalyptophis peronii, Aipysurus eydouxii, Astrotia stokesii, Enhydrina schistosa, Lapemis curtus, Laticauda colubrina, L. laticaudata and many Hydrophis species have been collected in both the Asian and the Australian region and are abundant in these areas. Other species are much less well known and some species are only collected from very restricted areas and therefore much more vulnerable for change in the environments. One of the rarest species is Hydrophis parviceps which is known from only two specimens collected in a very limited area in the southern Vietnamese part of the South China Sea. Hydrophis sibauensis is a recently described species which has been collected more than 150 km into rivers of Borneo and is only known from 3 specimens. This is also the case for another recently described species, Hydrophis laboutei which is only known from two specimens collected at Chesterfield Reefs (New Caledonia). Other species with a limited distribution are therefore highly vulnerable to all kinds of environmental changes, including Aipysurus fuscus, A. apraefrontalis and A. foliosquama which have only been collected in the north west part of Australian coral reefs. Recent surveys indicate that there has been a drastic decline of specimens in this area over the last 10–15 years and pers.comm. Michael Guinea). Hydrophis semperi and Laticauda crockeri, both species only known from Lake Taal in the Philippines and Lake Te-Nggano in Rennell Island, respectively, also represent species that have a very restricted distribution and therefore are highly vulnerable to changes in the environment.

Our own (ARR) investigations in the Gulf of Thailand and part of Borneo indicated that the sea snake fauna in this area has also declined. In the Gulf of Thailand, especially populations of species going up rivers have disappeared (e.g. Hydrophis torquatus and H. klossi), but also in Borneo the number of specimens collected previously in great numbers inside river mouths appears to have declined. Trawl fisheries in the Andaman Sea (e.g. around Phuket, Thailand) also got fewer specimens and the fishermen also mention that the size of sea snakes they got in former times was much larger than the ones they get now. A small survey in Cambodia (ARR) in the year 2000/2001 indicated similar results to those found for Borneo and Thailand.

What can we do to help the Sea Snakes to Survive?

The main threat to sea snakes is the cultural indifference to conservation issues by locals and, as a consequence, their commercial exploitation. Only raising awareness may reduce this kind of threat. Another problem in trying to help the sea snakes to survive is the very limited knowledge on their biology, especially in Asia. It is very important to get much more information on the biology in formulating management plans. For the moment we still miss information on breeding cycles, by-catch and mortality, growth rates, population density, sexual maturity and taxonomy in most areas. The effects of the exploitation and by-catch on the sea snakes are almost unknown except from the Philippines and Australia. Some populations may already be in danger of extinction. The only way to have a sustainable yield is by monitoring and controlling by-catch and commercial catch of sea snakes, giving local governments a chance to intervene before a catastrophic collapse of local populations occur. However, to limit exploitation of the most common species sustainable and to protect the endangered ones, we need to have much more biological focus on the group.

One cause for the disappearance of sea snake species from rivers is at least in part due to the great problems with the ongoing pollution in many rivers in both Asia and Australia. Also the information on breeding areas for most sea snake species is unknown, but there is no doubt that also habitat destructions (eg. mangrove clearings and water crafts) have a negative influence on sea snake populations.

A first step in sea snake conservation is to distinguish the many species from each other, which is not an easy task, and with the global decline of taxonomists it may be more important to focus initially on the entire group. The next step is to get more knowledge of sea snake biology and then to focus on the by-catch and management plan to protect the endangered species and harvest only the more common ones. Concerning the more global problems of pollution and habitat degradation, we have to put more pressure on politicians and hope that they will come up with solutions for the benefit of both humans and sea snakes.

Other Snakes found in Marine and Brackish Environments

Evolution is continually tinkering with snake populations that live in coastal areas adjacent to brackish and salt water environments. Besides the true sea snakes, many terrestrial and arboreal species have learned to exploit marine resources by foraging in the intertidal zone at low tide or from the branches of mangroves, while some freshwater species have adapted to life in brackish water, sometimes enter the ocean, or live there permanently.

All three living species of the ancient Acrochordidae use a combination of aquatic environments ranging from freshwater to sea water, and have probably been living in coastal ocean habitats for most of their 90.7 million year history. Acrochordids, particularly Acrochordus granulatus have some of the most specialized morphology and physiology for life in saltwater, including the greatest capacity to store oxygen found in any vertebrate. These are low metabolism snakes, feeding and reproducing infrequently and incapable of active swimming for more than a few minutes. Acrochordus arafurae females ambush prey in deep water while males actively search for prey in shallow water. They have a constriction-like behavior for holding prey between body loops during ingestion, and enlarged keels on their scales aid in holding slippery fish. Additionally each scale contains a mechanoreceptor that may be used to locate fish in turbid water.

The Asian and Australasian Homalopsidae, ironically sometimes called the "freshwater snakes," have species that live in coastal habitats such as mangrove forests and salt marshes. Homalopsids inhabit life zones that range from the fossorial Brachyorrhos albus and the semi-terrestrial Enhydris plumbea, to Erpeton tentaculatus that is all but incapable of leaving their freshwater habitats, to species inhabiting coastal marine environments (Bitia hydroides, Cantoria violacea, Myron sp., Fordonia leucobalia, Enhydris bennettii). The most widespread and successful brackish water homalopsids are the bockadams (Cerberus sp) which are distributed from the vicinity of Mumbai, India in the east to Palau, Micronesia in the west, and range southward into the Indonesian Archipelago, New Guinea, and northern Australia. Like some of the Hydrophis and Laticauda, the Lake Buhi Bockadam (Cerberus microlepis) is a freshwater species, derived from its nearby salt water dwelling relative, Cerberus rynchops in the Philippines. While most of the aquatic homalopids are piscivorous, three species, all members of the same clade specialize in feeding on crustaceans (Cantoria violacea, Fordonia leucobalia, and Gerarda prevostiana) in near shore habitats.

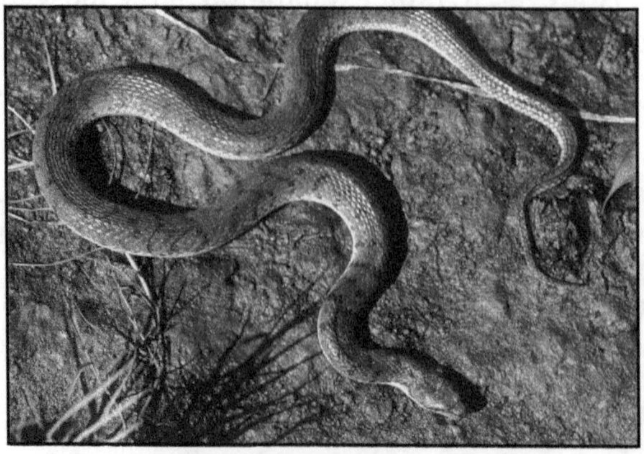

A homalopsid snake (Cerberus rynchops)

An assemblage of mangrove homalopsids studied in Singapore was dominated by the piscivorous Asian Bockadam, Cerberus rynchops (73% of total snakes), while the other species were crustacean specialists and less common. Gerard's Mud Snake, Gerarda prevostiana (16% of the total snakes) feeds exclusively on recently-molted crabs; while the Crab-eating Snake, Fordonia leucobalia, (10% of total snakes) specializes in feeding on hard-shell crustaceans. The most uncommon Singapore homalopsid was Cantor's Mud Snake, Cantoria violacea, (2% of total snakes) and it too has a specialized diet, feeding on Alpheus snapping shrimp (Decapoda, Alpheidae). The crustacean-eaters were often observed in association with mud lobster mounds constructed by Thalassina anomala (Decapoda: Thalassinidae). The snakes were nocturnal and active throughout the night. Gerard's Mud Snake increased activity during spring tides, but the other species did not. Three male Crab-eating snakes were tracked using radiotelemetry: they rarely moved and when they did it was for short distances (1.8 to 14.0 m). As might be expected for tropical aquatic snakes, the body temperatures were very stable (26.3–29.0 °C) and consistently higher than the microhabitat temperatures. Two of the radio tracked crab-eating Snakes made extensive use of mud lobster mounds.

Snakes of the family Natricidae (or subfamily Natricinae) make up about 30 genera and 210 species. They occur in both hemispheres and in both temperate and tropical environments, but they appear to have originated in Asia, and dispersed into Europe, and North America. Many of these snakes are semi-aquatic and some inhabit brackish water. In North Africa and Europe at least three species use coastal habitats: the Viperine Snake, Natrix maura, the Grass Snake, N. natrix sicula, and the Dice Snake, N. tessellata. North American brackish water natricids use mangrove and salt marsh habitats include the Mangrove Water Snake, Nerodia clarkii, in the southeastern USA, and the Baja Garter Snakes, Thamnophis valida. Both these species are piscivorous and like some freshwater natricids, Nerodia clarkii juveniles lure prey with their tongue. The Salt Marsh Brown Snake, Storeria dekayi limnectes is less aquatic than the previous species, and it feeds on soft bodied invertebrates.

In the Neotropics, the family Dipsadidae (or subfamily Dipsadinae) has more than 92 genera and >700 specie, a few of which use brackish and salt water habitats (Helicops, Hydrops, Liophis, and Tretanorhinus) to varying degrees. For the most part these snakes are poorly known and their brackish water habits are known only from anecdotal observations. Similarly, at least two aquatic African snakes, of uncertain lineages (Grayia smythii and Crotaphopeltis hotamboeia) occur in the brackish water of mangrove forests as well as freshwater.

There is another group of snakes that use marine resources that is worthy of note. These species forage in mangroves, salt marshes, and intertidal zone without actually spending much time in the water. The most specialized is perhaps the Burmese Vine Snake, Ahaetulla fronticincta (family Colubridae) which hunts gobies from branches over hanging the water. The Southeast Asian Mangrove Pitviper, Cryptelytrops purpureomaculatus (Viperidae) also hunts from the mangrove branches but its diet is unknown. There are others, for the most part habitat generalists with populations in or adjacent to marine habitats and include Python molurus, Python reticulatus (Pythonidae), Ophiophagus hannah (Elapidae), and Boiga dendrophilia (Colubridae) – all medium to large snakes that forage for food in coastal habitats. Other snakes forage in the intertidal zone: Coluber anthonyi (Colubridae); Crotalus mitchelli and Crotalus muertensis (family Viperidae, subfamily Crotalinae).

Thus, of 34 lineages of snakes (families and subfamilies), four (Acrochordidae, Homalopsidae,

Dipsadidae, and Elapidae) contain most of the species adapted for marine environments, while other clades have relatively few, or no, species adapted for the saline water.

Marine Iguana

The marine iguana is the only lizard in the world with the ability to live and forage at sea and is endemic to the Galapagos Archipelago. There are six very similar subspecies, each from different islands, with those from Isabela and Fernandina being the largest. The adults are black for most of the year, however the males change colour during the mating season – with different subspecies adopting different colourations. Near Española and Floreana they are most colourful – turning bright green and red; on Santa Cruz they are red and black; and on Fernandina they become dull green and brick red. The juveniles are generally black, with a lighter dorsal stripe than the adults. During the mating season males fight for dominance over harems of females, which they will fiercely defend from rival males.

Marine iguanas are not a very agile species on land, but they are excellent swimmers – moving easily through the water as they feed on algae. Larger individuals go further out to sea and use their powerful claws to grip on to rocks in strong currents to feed, whilst the smaller ones stay inshore near rock pools, feeding on algae exposed at low tide. Due to the high concentration of salt in their diet, marine iguanas filter their blood at the nose and sneeze out the excess salt, often forming salt crystals on the snout.

Despite not being truly social, iguanas are highly gregarious – especially on cold nights when they tend to group together to conserve heat. In the morning, they bask in the sun, absorbing heat with their black scales until they have enough energy to swim out to sea to forage. Upon entering the water, their heartbeat slows to half its normal pace in order to conserve energy and allow them to feed for as long as possible.

They can be seen in coastal regions on the islands of Isabela, Fernandina, Española, Floreana, Santa Cruz and other islets around them. They live on the islands all year round, with the mating season occurring between January and March. They are active during the day and spend a lot of time on rocks basking in the sun.

Threats - The introduction of cats and dogs by humans has greatly affected the numbers of marine iguanas as they predate upon the juveniles, which are poorly adapted to defending themselves against large land predators. El Niño periodically decreases the iguana population (by up to 85%), as it cuts short their food supplies. Oil spills may also have similarly dramatic effects on the population. They could also be threatened by marine plastic pollution – specifically microplastics.

Range and Habitat

Marine iguanas are the only seagoing lizards in the world, and they are found on the rocky shores of most of the Galapagos Islands. In fact, they range only in the Galapagos. The absence of mammalian predators has helped them adapt well to a marine environment. Because the water surrounding the islands can be quite cold, and they are ecothermic creatures, they come ashore often to warm up. They live in colonies near shallow reefs and rocky coastlines.

Behavior and Communication

Marine iguanas are most active during the day, that is, after spending the first few hours of the day sunbathing and preparing for the day's activities.

Feeding Habits

Marine iguanas are herbivores and feed mainly on intertidal seaweed and algae during low tide. When feeding, the larger iguanas, which can retain more body heat, may remain submerged in the water for up to one hour. Iguanas tend to intake a large amount of sodium, which can be toxic to them. To accommodate this, they excrete salt crystals by sneezing a lot.

Breeding and Reproduction

Breeding occurs at different times on different islands, but generally individuals breed once every two years. Mating season is January through April, depending on the island. They are colonial, often seen piling atop one another. When breeding, the larger males become territorial and aggressive, butting and pushing against rivals. They tend to want to conserve energy, so they may bite only to defend their territory.

Mated females lay one to six eggs in a sandy nest or volcanic ash burrow that is 1,000 feet inland. Females guard the nest fiercely for several days but then leave the eggs while they finish incubating, which takes approximately 95 days. Hatchlings are not given any parental protection.

Conservation

Marine iguanas are vulnerable, but since the creation of the Galapagos Marine Reserve in 1998, their status has somewhat improved. Their primary predators are birds, such as hawks, owls and herons when they are young, and the Galapagos hawk when they are adults. And they are also threatened by introduced cats, rats, dogs and pigs. In addition, El Niño events can be devastating, causing as many as 70 percent of the population to die in one year due to famine conditions.

Efforts made through the Charles Darwin Research Station to measure the iguanas' corticosterone levels—basically, adrenaline—helps monitor the general health of the marine iguana population. Their stress increases dramatacially during El Niño years and it decreases during La Niña when there is more food to feast on. It also increases when there is more human activity in their environment. When stress increases, there are many more complications with reproduction, such as eggs not being fertilized or few eggs being laid.

Crocodile

Crocodile is the common name for any species belonging to the reptile family Crocodylidae (order Crocodilia). The term also is used to refer to all members of the order Crocodilia, a group that includes the "true crocodiles" (family Crocodylidae), the alligators and caimans (family Alligatoridae), and the gharials (family Gavialidae), as well as the Crocodylomorpha, which includes prehistoric crocodile relatives and ancestors.

Crocodiles (both family and order) are large, primarily aquatic reptiles that are primarily found in tropical regions of Asia, the Americas, Africa, and Oceania, and occupy both freshwater and marine environments. They have a lizard-like form and swim and feed in the water, but breed on the land.

While advancing their own individual need for survival, maintenance, and reproduction, crocodiles also provide benefit for the ecosystem and for humans. Ecologically, they serve as apex predators, keeping prey populations in check. While the larger crocodiles can be very dangerous to human beings, crocodiles do provide commercial, nutritional, and aesthetic benefit. Their hide is tanned and used to make leather goods, and crocodile meat is considered a delicacy in many parts of the world. They also add to the human wonder and enjoyment of nature.

Order Crocodilia

Crocodilia is an order of large reptiles that appeared about 84 million years ago in the late Cretaceous period. The basic crocodilian body plan is a very successful one; modern species closely resemble their Cretaceous ancestors. Mammals, too, have adapted to this body plan at least once in history. The standard vernacular term for this group is "crocodilians" rather than "crocodiles," but the latter term also is used in reference to this group. Crocodilian will be the term used here to refer to the members of the order.

Crocodilians are considered the closest living relatives of birds, as the two groups are believed to be the only survivors of the Archosauria. Members of the crocodilian stem group, the clade Crurotarsi, appeared about 220 million years ago in the Triassic period and exhibited a wide diversity of forms during the Mesozoic era.

The group is often spelled Crocodylia for consistency with the genus Crocodylus Laurenti, 1768. However, Richard Owen used the -i- spelling when he published the name in 1842, so it is generally preferred in the scientific literature. The -i- spelling is also a more accurate Latinization of the Greek (krokodeilos, literally "pebble-worm," referring to the shape and texture of the animal). The crocodile gets its name from the Greeks who observed them in the Nile River. The Greeks called them krokodilos, a compound word from kroke, which means "pebbles" and drilos, which means "worm." To the Greeks, this "worm of the stones" was so named because of the crocodiles' habit of basking in the sun on gravel-covered river banks.

Crocodilians have a flexible semi-erect (semi-sprawled) posture. They can walk in low, sprawled "belly walk," or hold their legs more directly underneath them to perform the "high walk" . Most other reptiles can only walk in a sprawled position, and chameleons are the only modern reptiles with a more erect posture than crocodilians. The semi-erect posture makes it possible for some species to gallop on land if necessary. An Australian species can reach a speed of over 16 km/h while galloping on an irregular forest floor. Their ankle bones or tarsi are highly modified. Modern crocodilian locomotion is not a primitive trait, but a specialization for their semi-aquatic lifestyle.

All crocodilians have, like *Homo sapiens* (humans), thecodont dentition (teeth set in bony sockets), but unlike mammals, they replace their teeth throughout life (though not in "extreme" old age). Juvenile crocodilians replace teeth with larger ones at a rate as high as 1 new tooth per socket every month. After reaching adult size in a few years, however, tooth replacement rates can slow to two years and even longer. Very old members of some species have been seen in an almost "edentulous" (toothless) state, after teeth have been broken and replacement slowed or ceased. The result of this is that a single crocodile can go through at least 3,000 teeth in its lifetime. Each tooth is hollow, and the new one is growing inside the old. In this way, a new tooth is ready once the old is lost.

From the top: Head of an American alligator (Alligator mississippiensis), a Nile crocodile (Crocodylus niloticus), and an Indian gharial (Gavialis gangeticus)

Crocodilians have a secondary bony palate that enables them to breathe when partially submerged, even if the mouth is full of water. Their internal nostrils open in the back of their throat, where a special part of the tongue called the "palatal valve" closes off their respiratory system when they are underwater. This way they can open their mouths underwater without choking. Most reptiles lack a secondary palate, but some skinks (family Scincidae) have a bony secondary palate too, to varying degrees.

Crocodiles and gharials have modified salivary glands on their tongue (salt glands), which are used for excreting excess salt ions from their body. Alligators and caimans have them too, but here they are non-functioning. This indicates that at some point the common origin of the Crocodylia was adapted to saline/marine environments. This also explains their wide distribution across the continents (i.e. marine dispersal). Species like the saltwater crocodile (*C. porosus*) can survive protracted periods of time in the sea, and can hunt prey within this environment.

Crocodilians are often seen lying with their mouths open, a behavior called gaping. One of its functions is probably to cool them down, but since they also do this at night and when it is raining, it is possible that gaping has a social function too.

Internal Organs

Crocodilians lack a vomeronasal organ (yet it is detectable in the embryo) and a urinary bladder. Like mammals and birds, and unlike other reptiles, crocodiles have a four-chambered heart; however, unlike mammals, oxygenated and deoxygenated blood can be mixed because of the presence of the left aortic arch. The right ventricle has two arteries leaving it; a pulmonary artery, which goes to the lungs, and the left aortic arch, which goes to the body, or systemic circulation. There is also a hole, the foramen of Panizza, between the left and right aortic arches. Because the left aortic arch goes directly to the gut, the shunting of oxygen-depleted blood which is high in CO_2 may serve to aid in creating stomach acid to assist in digesting bones from its prey.

They have alveoli in their lungs and a unique muscular attachment to the liver and viscera that acts as a piston to breathing, separating the thoracic and abdominal cavities (similar to the diaphragm of mammals). Although tegu lizards have a primitive proto-diaphragm, separating the pulmonary cavity from the visceral cavity and allowing greater lung inflation, this is considered to have a different evolutionary history.

Crocodilians are known to swallow stones, gastroliths ("stomach-stones"), which act as a ballast in addition to aiding post-digestion processing of their prey. The crocodilian stomach is divided into two chambers, the first one is described as being powerful and muscular, like a bird gizzard. This is where the gastroliths are found. The other stomach has the most acidic digestive system of any animal, and it can digest mostly everything from their prey; including bones, feathers, and horns.

The gender of the juvenile is determined by the incubation temperature. This means crocodilians do not have genetic sex determination (like us), but a form of environmental sex determination that is based upon temperature embryos undergo early in their development.

Sensory Organs

Like all reptiles, crocodilians have a relatively small brain, but it is more advanced than in other reptiles. Among other things it has a true cerebral cortex.

As in many other aquatic or amphibian tetrapods, the eyes, ears, and nostrils are all located on the same plane. They see well during the day and may even have color vision, plus the eyes have a vertical, cat-like pupil that also gives them excellent night vision. The iris is silvery (light-reflecting layer of tapetum behind the retina greatly increases their ability to see in weak light), making their eyes glow in the dark. A third transparent eyelid, the nictitating membrane, protects their eyes underwater. However, they cannot focus underwater, meaning other senses are more important when submerged underwater.

While birds and most reptiles have a ring of bones around each eye that supports the eyeball (the sclerotic ring), the crocodiles lack these bones, just like mammals and snakes. The eardrums are located behind the eyes and are covered by a movable flap of skin. This flap closes, along with the nostrils and eyes, when they dive, preventing water from entering their external head openings. The middle ear cavity has a complex of bony air-filled passages and a branching eustachian tube. There is also a small muscle (which is also seen in geckos) next to or upon the stapes, the stapedius, which probably functions in the same way as the mammalian stapedius muscle does, dampening strong vibrations.

The upper and lower jaws are covered with sensory pits, seen as small, black speckles on the skin, the crocodile version of the lateral organ seen in fish and many amphibians. But they have a completely different origin. These pigmented nodules encase bundles of nerve fibers that respond to the slightest disturbance in surface water, detecting vibrations and small pressure changes in water, making it possible for them to detect prey, danger, and intruders even in total darkness. These sense organs are known as DPRs (Dermal Pressure Receptors). While alligators and caimans only have them on their jaws, crocodiles have similar organs on almost every scale on their body. The function of the DPRs on the jaws is clear, but it is still not quite clear what the organs on the rest of the body in crocodiles actually do. They are probably doing the same as the organs on their jaws, but it seems like they can do more than that, like assisting in chemical reception or even salinity detection.

Skin and Skeleton

West African dwarf crocodile from the forests of West and West Central Africa

The skin is covered with non-overlapping scales composed of the protein keratin (the same protein that forms hoofs, skin, horns, feathers, hair, claws, and nails in other tetrapods), which are shed individually. On the head, the skin is actually fused to the bones of the skull. There are small plates of bone, called osteoderms or scutes, under the scales. Just like a tree, crocodile osteoderms have annual growth rings, and by counting them it is possible to tell their age. Osteoderms are found especially on the back, and in some species also on the belly. The overlapping rows of scutes cover the crocodile's body from head to tail, forming a tough protective armor. Beneath the scales and osteoderms is another layer of armor, both strong and flexible and built of rows of bony overlapping shingles called osteoscutes, which are embedded in the animal's back tissue. The blood-rich bumpy scales seen on their backs act as solar panels.

The spool-shaped vertebrae in their ancestors went from being biconcave to having a concave front and a convex back in the modern forms. This made the vertebral column more flexible and strong, a useful adaptation when hunting in water. They possess ribs of dermal origin restricted to the sides of the ventral body wall. The collar bone (clavicle) is absent.

Family Crocodylidae

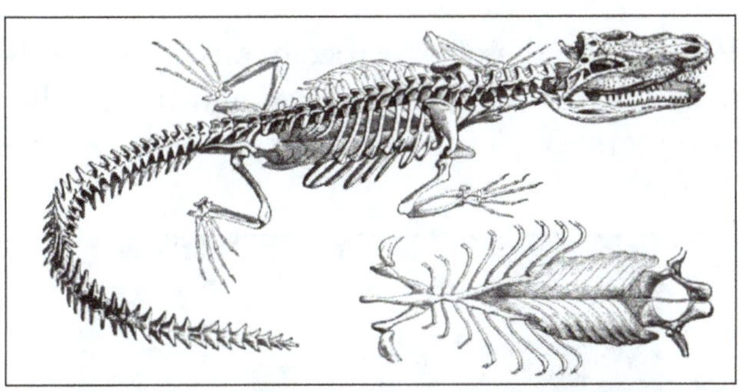

Crocodiles, like dinosaurs, have the abdominal ribs modified into gastralia

The family Crocodylidae comprises the "true crocodiles." These live throughout the tropics in Africa, Asia, the Americas, and Australia. Crocodiles tend to congregate in freshwater habitats like rivers, lakes, wetlands, and sometimes in brackish water. Some species, notably the saltwater crocodile of Australia, Southeast Asia, and the Pacific islands often live along the coastal areas. It is also known to venture far out to sea.

Crocodiles are more closely related to birds and dinosaurs than to most animals classified as reptiles, the three being included in the group Archosauria ("ruling reptiles"). Crocodiles mostly feed on vertebrates like fish, reptiles, and mammals, and sometimes on invertebrates like mollusks and crustaceans, depending on species.

Their external morphology is a sign of their aquatic and predatory lifestyle. A crocodile's physical traits allow it to be a successful predator. They have a streamlined body that enables them to swim faster. They also tuck their feet to their sides while swimming, which makes the animal even faster, by decreasing the water resistance. They have webbed feet which, although not used to propel the animal through the water, allow it to make fast turns and sudden moves in the water or initiate swimming. Webbed feet are an advantage in shallower water where the animals sometimes move around by walking.

While often confused with each other, alligators and crocodiles belong to two quite separate taxonomic families, and are as distinct from one another as humans are from gorillas. As for appearance, one generally reliable rule is that alligators have U-shaped heads, while crocodiles are V-shaped—which can be remembered by noting that "A" in alligator comes before "C" in crocodile, and "U" comes before "V."

Crocodiles are very fast over short distances, even out of water. They have extremely powerful jaws capable of biting down with 3,000 pounds of pressure per square inch, and sharp teeth for tearing flesh, but cannot easily open their mouth if it is held closed. There are stories of people escaping from the long-snouted Nile crocodile by holding its jaws shut. Zoologists will often subdue crocodiles for study or transport by taping their jaws or holding their jaws shut with large rubber bands cut from automobile inner tubes. All large crocodiles also have sharp and powerful claws. They have limited lateral movement in their neck, so on land protection can be found by getting even a small tree between the crocodile's jaws and oneself.

Age and Size

There is no reliable way of measuring crocodile age, although several techniques could be used to derive a reasonable guess. The most common method is to measure lamellar growth rings in bones and teeth—each ring corresponds to a change in growth rate, which typically occurs once a year between dry and wet seasons.

Bearing these inaccuracies in mind, the oldest crocodilians appear to be the largest species. *porosus* is estimated to live around 70 years on average, and there is limited evidence that some individuals may exceed 100 years. One of the oldest crocodiles recorded died in a zoo in Russia apparently aged 115 years old.

Large crocodile in captivity in Australia

Size greatly varies between species, from the dwarf crocodile to the enormous saltwater crocodile. Large species can reach over 5 or 6 meters long and weigh well over 1200 kilograms (2,640 pounds). Despite their large adult size, crocodiles start their life at around 20 centimeters long. The largest species of crocodile is the saltwater crocodile, found in northern Australia and throughout Southeast Asia.

Biology and Behavior

Crocodile farm in Mexico

Crocodiles eat fish, birds, mammals, and occasionally smaller crocodiles. Crocodiles are ambush hunters, waiting for fish or land animals to come close, then rushing out to attack. As cold-blooded predators, they can survive long periods without food, and rarely need to actively go hunting. The crocodile's bite strength is up to 3,000 pounds per square inch (psi), comparing to just 100 psi for a Labrador retriever, 350 psi for a large shark, or 800 psi for a hyena. Despite their slow appearance, crocodiles are top predators in their environment, and various species have been observed attacking and killing sharks. A famous exception is the Egyptian plover, a bird which is said to enjoy a mutually beneficial symbiotic relationship with the crocodile. According to reports, the plover feeds on parasites that infest the crocodile's mouth and the reptile will open its jaws and allow the bird to enter to clean out the mouth.

Crocodiles and Humans

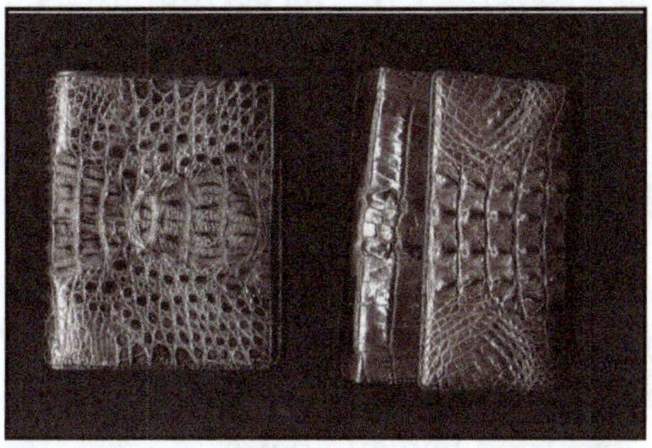

Crocodile leather wallets from Bangkok Crocodile Farm

Wild crocodiles are protected in many parts of the world, but they also are farmed commercially. Their hide is tanned and used to make leather goods, such as shoes, wallets, briefcases, belts, hats, and handbags. Crocodile meat is also considered a delicacy in many parts of the world. The most commonly farmed species are the saltwater and Nile crocodiles, while a hybrid of the saltwater and the rare Siamese crocodile is also bred in Asian farms. Farming has resulted in an increase in the saltwater crocodile population in Australia, as eggs are usually harvested from the wild, so landowners have an incentive to conserve crocodile habitat.

Crocodile meat is consumed in some countries, such as Australia, Ethiopia, Thailand, South Africa, and also Cuba (in pickled form). It can also be found in specialty restaurants in some parts of the United States. The meat is white and its nutritional composition compares favorably with that of more traditional meats. It does tend to have a slightly higher cholesterol level than other meats. Crocodile meat has a delicate flavor and its taste can be complemented by the use of marinades. Choice cuts of meat include back strap and tail fillet.

The larger species of crocodiles can be very dangerous to humans. Crocodiles are the leading cause of animal-related deaths as of 2001. The saltwater and Nile crocodiles are the most dangerous, killing hundreds of people each year in parts of Southeast Asia and Africa. Mugger crocodiles and possibly the endangered black caiman also are very dangerous to humans. American alligators are less aggressive and rarely assault humans without provocation.

The most deaths in a single crocodile attack incident may have occurred during the Battle of Ramree Island, on February 19, 1945, in what is now Myanmar. Nine hundred soldiers of an Imperial Japanese Army unit, in an attempt to retreat from the Royal Navy and rejoin a larger battalion of the Japanese infantry, crossed through ten miles of mangrove swamps that contained saltwater crocodiles. Twenty Japanese soldiers were captured alive by the British, and almost five hundred are known to have escaped Ramree. Many of the remainder may have been eaten by the crocodiles, although gunfire from the British troops was undoubtedly a contributory factor.

Marine Birds

'Seabird' is a general term used to describe collectively any bird species which spends a substantial part of its life foraging and breeding in the marine environment. Birds considered being seabirds include penguins, albatrosses, petrels, shearwaters, cormorants, gannets, boobies, gulls and terns. Some species of pelicans, ducks and grebes also may forage at sea at times but are not generally considered to be seabirds.

Biology, Diet and Breeding

Seabirds spend much of their lives roaming the oceans in search of prey. Most species tend to forage on their own, though large feeding flocks will gather at rich or passing food sources. Squid, fish and krill are common sources of food. Many species will follow fishing vessels, aggressively competing with each other for discarded fish and baits.

Many seabirds, such as albatrosses and petrels, have long life spans with a few individuals known to live for more than 60 years. They have low rates of natural mortality. Albatrosses and petrels achieve sexual maturity at 5-12 years of age and breed in colonies on remote islands, with the pair bond being reinforced by elaborate courtship displays. Most species attempt to breed annually, although the larger albatrosses breed successfully only every second year (biennially).

Parental duties are shared by both sexes. Petrels and shearwaters may variously nest in simple scrapes on the ground, on cliff ledges, in a burrow or in a rocky crevice. Albatrosses and giant petrels breed on the surface and, where nest material is available, build large bowl-shaped nests of grass and mud. A single, large (relative to body size), white egg is laid which both parents incubate for 35-85 days in alternating shifts. After hatching, chicks are brooded for a short period until they are able to regulate their own body temperature. After this period the chick is generally left alone, with parents returning to provide food at intervals. Both parents feed the chick until it fledges at 110-304 days, depending on the species.

Habitat

Seabirds occur widely across the world's oceans. Nineteen of the World's 22 albatross species occur in the Southern Hemisphere. Many species, such as Grey-headed Albatrosses, are extremely dispersive, spending most of their time over the surface waters of the high seas. In contrast, others, like adult Shy Albatrosses, tend to be more sedentary, regularly foraging over coastal waters throughout their adult lives.

Threats

Past threats to seabirds include being killed for meat, eggs and feathers. Modification of breeding habitats and introduced predators, such as feral cats, also impact considerably on populations. More recently, the global expansion of longline fisheries has posed one of the greatest overall threat to seabirds, although trawling and gillnetting also has led to the death of many albatrosses and petrels.

Longline fishing is a method used to target finfish and shark species. A longline consists of a main line with numerous baited hooks attached to branch lines. The line can be set relatively close to the sea surface (pelagic) or on the seabed (demersal). The number and type of hooks and the length of the branch lines depend on the target species. Pelagic longlines can be up to 100 km long and carry a few thousand hooks. However, demersal longlines, although shorter in length, can hold more than 20 000 hooks. Each year thousands of seabirds are accidentally killed when birds ingest baited hooks during the setting or hauling of the line. Birds hooked during setting are subsequently pulled under the water by the weight of the longline and drown. The level of mortality that occurs in longline fisheries is not sustainable for many populations of seabirds.

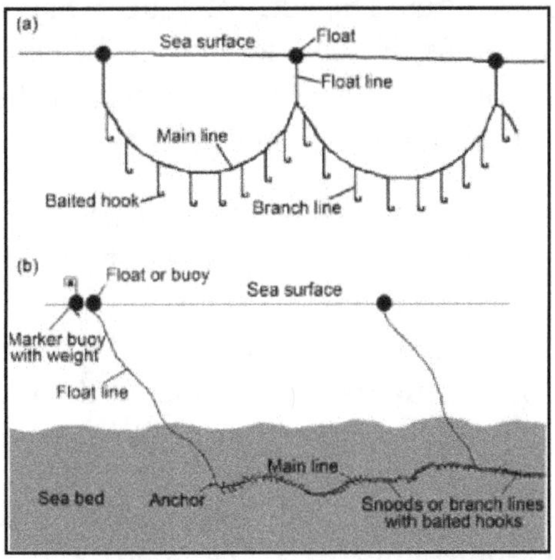

Longline fishing: (a) Pelagic (b) Demersal

However, this threat can be greatly minimised by modifying fishing practice and adopting mitigation measures. These include the use of bird-scaring lines, weighted lines to reduce the amount of time baits are available to birds at or near the surface, setting lines only at night and seasonal closures of fisheries to avoid fishing when birds are more susceptible to being caught, such as around colonies during the breeding season. Adoption of these measures has now virtually eliminated seabird by catch in some fisheries.

Auk

Auk is the common name for any member of the bird family Alcidae in the order Charadriiformes. They are diving seabirds with a chunky body, short wings and legs, and webbed feet and are found in northern climates (Arctic, subarctic, temperate regions). The about two dozen recognized species of extant (living) auks are placed in the subfamily Alcinae and include true auks and murres (tribe Alcini), murrelets and guillemots (tribes Brachyamphini, Synthliboramphini, and Cepphini), puffins (tribe Fraterculini), and auklets (tribe Aethiini).

Auk's wing-propelled diving ability allows them to be efficient for capturing certain prey, depending on their particular adaptations. Some, such as the murres *(Uria),* are more efficient under water and can capture faster schooling fish; others, such as the auklets, which are better adapted for flying and walking, tend to take slower moving krill. In addition to playing key roles in marine food chains, auks, with their unique adaptations, including breeding behavior, add to the human fascination with nature. Despite these values, auks have sometimes suffered greatly from anthropogenic factors. The great auk was hunted on such as significant scale for food, eggs, and down that it became extinct by the mid-nineteenth century.

Description and Distribution

The auks are restricted to cooler northern waters, with most species in the Pacific. Their ability to spread further south likely is restricted as their prey hunting method, pursuit diving, becomes less efficient in warmer waters. The speed at which small fish (which along with krill are the auk's principal food items) can swim doubles as the temperature increases from 5 °C to 15 °C, with no corresponding increase in speed for the bird. The southernmost auks, in California and Mexico, can survive there because of cold upwellings. The current paucity of auks in the Atlantic (six species), compared to the Pacific (19-20 species) is considered to be because of extinctions to the Atlantic auks; the fossil record shows there were many more species in the Atlantic during the Pliocene. Auks also tend to be restricted to continental shelf waters and breed on few oceanic islands.

Many species of auks, principally the true auks, are superficially similar to penguins due to their black-and-white colors, their upright posture, and some of their habits. Nevertheless they are not related to the penguins at all, but considered by some to be a product of moderate convergent evolution. In contrast to penguins, modern auks are able to fly (with the exception of the recently extinct Great Auk). They are good swimmers and divers, but their walking appears clumsy. Due to their short wings auks have to flap their wings very quickly in order to fly.

Extant auks range in size from the least auklet *(Aethia pusilla),* at 85 grams (3 ounces) and 15 centimeters (six inches), to the thick-billed murre *(Uria lomvia),* at one kilogram (2.2 pounds) and 45 cm (18 inches) and the razonbill *(Alca torda)* at up to 43 cm in length with a 60-69 cm wingspan.

The flightless Great Auk *(Pinguinus impennis)* was the largest, standing about 75 centimeters (30-34 inches) high and weighing around five kilograms, but it became extinct around the mid-nineteenth century.

Several species have different names in Europe and North America. The guillemots of Europe are murres in North America, if they occur in both continents, and the little auk becomes the Dovekie. Auks live on the open sea and only go ashore for breeding, although some species, like the common guillemot, spend a great part of the year defending their nesting spot from others.

Some species, such as the *Uria* guillemots, nest in large colonies on cliff edges; others, like the *Cepphus* guillemots, breed in small groups on rocky coasts; and the puffins, auklets and some murrelets nest in burrows. All species except the *Brachyramphus* murrelets are colonial. Colonial species tend to use the same breeding grounds each year, and individually even the same nesting site.

Feeding and Ecology

The feeding behavior of auks is often compared to that of penguins; they are both wing-propelled pursuit divers. In the region where auks live, their only seabird competition is with cormorants (which dive powered by their strong feet); in areas where the two groups feed on the same prey the auks tend to feed further offshore.

Although not to the extent of penguins, auks have to a large extent sacrificed flight, and also mobility on land, in exchange for swimming; their wings are a compromise between the best possible design for diving and the bare minimum needed for flying. This varies by subfamily. The *Uria* guillemots (including the razorbill) and murrelets are the most efficient under the water, whereas the puffins and auklets are better adapted for flying and walking. This reflects the type of prey taken; murres hunt faster schooling fish, whereas auklets take slower moving krill. Time depth recorders on auks have shown that they can dive as deep as 100 meters in the case of *Uria* guillemots, 40 meters for the *Cepphus* guillemots, and 30 meters for the auklets.

Traditionally, the auks were believed to be one of the earliest distinct charadriiform lineages due to their characteristic morphology. However, molecular analyses have cast doubt on this view. For example, as opposed to plovers (a much older charadriiform lineage), auks radically changed from a wading shorebird to a diving seabird lifestyle. Thus, today, the auks are no longer separated in their own suborder (Alcae), but are considered part of the Lari suborder that otherwise contains gulls and similar birds. Judging from molecular data, their closest living relatives appear to be the skuas, with these two lineages separating about 30 million years ago. This may or may not be correct due to uncertainties of the fossil record. Alternatively, auks may have split off far earlier from the rest of the Lari and undergone strong morphological, but slow molecular evolution, which would require a very high evolutionary pressure, coupled with a long lifespan and slow reproduction.

The earliest unequivocal fossils of auks are from the Miocene (e.g. the genus *Miocepphus*, 15 million years ago). Two very fragmentary fossils often are assigned to the Alcidae, although this may not be correct: *Hydrotherikornis* (Late Eocene, some 35 million years ago) and *Petralca* (Late Oligocene). Most extant genera are known to exist since the Late Miocene or Early Pliocene (about 5 million years ago). Miocene fossils have been found in both California and Maryland, but the greater diversity of fossils and tribes in the Pacific leads most scientists to conclude that it was there they first evolved, and it is in the Miocene Pacific that the first fossils of extant genera are found. Early movement between the Pacific and the Atlantic probably happened to the south (since there was no northern opening to the Atlantic), later movements across the Arctic Sea. The flightless subfamily Mancallinae which was apparently restricted to the Pacific coast of southern North America became extinct in the Early Pleistocene.

Razorbills are true auks only found in the Atlantic Ocean

The extant auks (subfamily Alcinae) are broken up into two main groups: the usually high-billed puffins (tribe Fraterculini) and auklets (tribe Aethiini), as opposed to the more slender-billed murres and true auks (tribe Alcini), and the murrelets and guillemots (tribes Brachyramphini and Cepphini). The tribal arrangement was originally based on analyses of morphology and ecology . mtDNA cytochrome *b* sequence, and allozyme studies confirm these findings except that the *Synthliboramphus* murrelets should be split into a distinct tribe, as they appear more closely related to the Alcini—in any case, assumption of a closer relationship between the former and the true guillemots was only weakly supported by Strauch's 1985 study.

Compared to other families of seabirds, there are no genera with many species (such as the 47 *Larus* gulls). This is probably a product of the rather small geographic range of the family (the most limited of any seabird family), and the periods of glacial advance and retreat that have kept the populations on the move in a narrow band of subarctic ocean.

Petrel

Petrel is any of a number of seabirds of the order Procellariiformes, particularly certain members of the family Procellariidae, which also includes the fulmars and the shearwaters. Members of the family Hydrobatidae are increasingly called storm petrels; those of the Pelecanoididae are usually called diving petrels.

Among the procellariid petrels, some two dozen species of the genera Pterodroma and Bulweria are called gadfly petrels because their flight is more fluttering than that of the related shearwaters. Certain heavy-bodied petrels are known as fulmars, and one, Macronectes giganteus, is called both giant fulmar and giant petrel.

Gadfly petrels nest in loose colonies on islands in the tropical and subtropical regions of the major oceans. A single egg (rarely two) is laid on the soil surface or in a burrow or crevice. The chick is tended by both parents and deserted about a week before it is fully fledged; it completes its development on stored fat. During the nonbreeding season, these birds roam the open ocean, feeding on squid and small fish. Most gadfly petrels are dark above and light beneath, with long wings and short, wedge-shaped tails. Because they are quite similar in appearance, the species are difficult to distinguish.

Fiji petrel (Pseudobulweria macgillivrayi)

Some of the better known gadfly petrels are the endangered Bermuda petrel, or cahow (Pterodroma cahow, sometimes considered a race of P. hasitata); the dark-rumped petrel, also called the Hawaiian petrel (P. phaeopygia), another endangered species, now concentrated almost entirely on the island of Maui; the phoenix petrel (P. alba), which breeds on several tropical archipelagos; and the black-capped petrel, or diablotin (P. hasitata), formerly found throughout the West Indies but now known only in Haiti. Several other species, including the Chatham Island (or magenta) petrel, the Galápagos dark-rumped petrel, and the Réunion petrel, are also on the endangered list.

Bonin petrel: Midway Atoll National Wildlife Refuge

Bonin petrel (*Pterodroma hypoleuca*) buried in the sand after a tsunami generated by an earthquake off Japan's coast in 2011 struck the Midway Atoll National Wildlife Refuge. It was later rescued.

Several other procellariids are also called petrels. Among them are the pintado petrel, or Cape pigeon (Daption capensis), a sub-Antarctic species about 40 cm (16 inches) long, marked with bold patches of black and white. The snow petrel (Pagodroma nivea), 35 cm, a pure white species, and the Antarctic petrel (Thalassoica antarctica), 42 cm, a brown-and-white-pied species, are rarely seen outside Antarctic waters.

Snow Petrel

The snow petrel is the smallest of the seven fulmarine petrel species, the other six being southern and northern giant petrels, Antarctic and northern fulmar, Antarctic petrel and Cape petrel. It is the only petrel species with pure white plumage, providing it with good camouflage in Antarctica. In contrast, it has a short, narrow, hooked blue-black bill, dark brown eyes and blue-black feet. It has relatively long wings and a wedge-shaped tail.

Along with South Polar skua, the snow petrel has the southernmost breeding range of any bird species, and is one of only three species recorded at the South Pole. It is a circumpolar breeder in Antarctica, nesting up to 440 km inland in the Antarctic mountains, and up to 2,400 m above sea level. It also breeds on glaciated islands, including the Balleny Islands and Scott Islands in the Ross Dependency, and South Georgia, South Orkney and the South Sandwich Islands, and Bouvetøya Island, in the South Atlantic Ocean. Its life-cycle is closely linked to pack ice: drifting sea ice that becomes packed together in large masses. Birds occur mainly in areas of sea with 10-50% ice cover, are rarely recorded at sea north of 55 °S, and have never been recorded around the New Zealand mainland.

Identification

The snow petrel is a small fulmarine petrel with pure white plumage, a short narrow hooked blue-black bill, dark eyes and bluish-black feet. It has relatively long wings (75-95 cm), and a wedge-shaped tail when fanned in flight. The sexes are alike, with the male usually slightly larger; juveniles are distinguished by extensive black barring on upperparts. Snow petrels are agile fliers, flying low when over water, using more rapid wing-beats and less gliding than other petrels. Pronounced differences in size are reported between individuals, with the largest adult birds around one quarter larger than the smallest.

- Voice: Adults cluck, screech and chirr. They have harsh gutteral calls, including a high pitched *kir-kir-kir*, and make plaintive calls during mating. Chicks *chirrup* when parents arrive at the nest.
- Similar species: The only other all-white bird in Antarctica is the snowy sheathbill (absent from the Ross Sea sector), which is distinguished by its bulkier body, larger head and yellow bill, and land-based life-cycle. Antarctic and Arctic terns in Antarctic seas are distinguished by black caps and red bills. White morph southern giant petrels are at least 8 x the size of the largest snow petrels (by weight), with huge pale yellowish-horn bills, and always have a few scattered black feathers.

Distribution and Habitat

Snow petrels are confined to Antarctica and neighbouring glaciated islands, including the Balleny Islands and Scott Islands in the Ross Dependency, and South Georgia, South Orkney and the South Sandwich Islands, and Bouvetøya Island, in the South Atlantic Ocean. In October 1994, two birds were recorded as vagrants at sea 180 nautical miles south-west of Australian-administered Macquarie Island. In 1994 and 1995, two dead birds were recovered on Macquarie Island, which lies 650-700 km south-west of Auckland and Campbell Islands in the New Zealand subantarctic region.

The snow petrel has a circumpolar distribution in Antarctic waters, from the edge of the pack-ice to the continental ice-shelf. Occasionally birds flock in open water where icebergs are present, where they are often seen roosting on icebergs. During the austral winter they disperse to pack ice, ice floes and open seas, when seas nearest breeding sites are frozen. Birds recorded at South Georgia are always near drifting ice offshore or at the heads of glaciers.

Population

There is an estimated total global population of 4 million adult snow petrels.

Breeding

The snow petrel is a circumpolar, loosely colonial annual breeder south of the Antarctic Convergence, on glaciated islands and continental Antarctica. Colonies in the Theron Mountains are up to 440 km inland from the coast, and at altitudes up to 2,400 m above sea level. It nests where rock is exposed during summer, under overhangs, or in crevices and small caves. Birds perform aerial displays year-round. Courtship involves head-raising, bill-fencing, bill "kissing", and mutual preening. Some birds stay at colonies year-round, with the main influx mid-September to early November. Snow petrels are monogamous with long-lasting pair bonds, and shared incubation and chick care. A single white egg (56 × 40 mm) is laid between late November and mid-December. Nests are pebble-lined scrapes on snow-free rock ledges with overhanging protection or in deep rock crevices. Incubation takes 41-49 days, with eggs hatching in January. Chicks fledge between late February and mid-May, 41-45 days after hatching.

Behaviour and Ecology

Snow petrels are loosely gregarious at sea and when breeding. Flocks show no coordinated flying movements; individuals have agile, erratic, bat-like flight. Snow petrels nesting south of the Antarctic Circle experience continuous daylight in the middle of their breeding cycle; even then, they concentrate their activity at the breeding colonies when the sun is lowest in the sky. Snow petrels are not truly colonial because nests are often sparsely scattered among large concentrations in suitable habitat. Nests are often associated with those of other fulmarine petrels (Antarctic fulmar, Antarctic petrel, Cape petrel), which compete for the same limited ice-free space. The snow petrel is more of a crevice nester than the three larger ledge-nesting species. At the nest, an occupying bird may squirt foul-smelling orange stomach oils at intruders (other snow petrels, skuas or humans) via the open mouth, and fight off other snow petrels using its bill and wings. Birds stained with oil from such fighting gather on snow banks near nesting sites to clean their feathers in snow. Snow petrels may live up to 20 years.

Food

Snow petrel's feeds mainly on krill, squid and fish, and scavenge carrion such as bird, seal and whale carcasses. Food is taken by aerial and contact dipping, surface-diving, surface-seizing and pursuit diving, and sometimes pursuit plunging with open wings. While surface-seizing, birds sometimes peck the sea surface continuously while rotating their bodies with their webbed feet, like a phalarope. Feeding birds sometimes accompany minke whales and Antarctic petrels.

Threats and Conservation

One of the most abundant bird species of the Antarctic region, the IUCN Red List classifies the snow petrel as Least Concern. Climate change may threaten the species in future because Antarctic birds seem to be breeding later in response to a reduction in the extent of sea ice cover, causing decreases in prey species in some areas, or the increasing length of the sea ice season, which delays access to breeding grounds.

Storm Petrel

Storm petrel is any member of about 20 species of seabirds constituting the family Hydrobatidae, or sometimes considered as Oceanitidae (order Procellariiformes). Ranging in length from about 13 to 25 centimetres (5 1/2 to 10 inches), all are dark gray or brown, sometimes lighter below, and often with a white rump. The wings are shorter than those of most other procellariiforms and are rounded at the tips; the bill is of medium length; the nostril tubes (common to all members of the order) are united on the upper surface of the bill; the toes are webbed, except for the hind toe, which is small and elevated; the tail is medium to long and square, forked, or wedge shaped.

Storm petrels fall into two behavioral and structural groups. Most of the species breeding in the southern oceans is shorter winged, square tailed, long legged, and short toed. With wings spread, they patter over the water, "walking," and pick up minute marine organisms. An example is Wilson's petrel (Oceanites oceanicus), which breeds on islets along the Antarctic continent and near the Antarctic Circle and winters in the North Atlantic from about June to September.

Most of the northern species have longer wings, forked or wedge-shaped tails, shorter legs, and longer toes; when foraging, they swoop over the water like tiny terns, occasionally alighting on the surface. Leach's petrel (Oceanodroma leucorhoa), for example, breeds on islands in the North Atlantic and south to about 28° N in the Pacific. Several other Oceanodroma species occur in the North Pacific. The British storm petrel (Hydrobates pelagicus) breeds on islands and cliffs along the coasts of Europe.

Storm petrels are colonial breeders, nesting in burrows they dug in soft soil or in crevices in rocks. They breed on islands along the coasts of North and South America, in the eastern and southern Atlantic, and throughout the Pacific. A single whitish egg is incubated by both parents, which come and go by night and so are rarely seen by man and other potential predators. Like many exclusively nocturnal birds, petrels are highly vocal at the colony. Like most other procellariiform young, the chicks are deserted by their parents a week or more before they are fully fledged. While the parents molt at sea, the young complete their growth and leave the burrow to feed them at sea.

Albatross

Albatrosses are large seabirds in the biological family Diomedeidae of the order Procellariiformes (the tubenoses). Albatrosses are among the largest of flying birds, and the great albatrosses (genus

Diomedea) have the largest wingspans of any extant (living) birds. They are closely related to the procellariids, storm-petrels, and diving-petrels, all of which also are part of the Procellariiformes. Some systematists recognize another order, Ciconiiformes, instead of Procellariiformes.

Albatrosses range widely in the Southern Ocean (South Pole Ocean or Antarctic Ocean) and the North Pacific Ocean. They are generally absent from the North Atlantic Ocean, although fossil remains show they once occurred there too and occasional vagrants are encountered.

Albatrosses are colonial, nesting for the most part on remote oceanic islands, often with several species nesting together. Pair bonds between males and females form over several years, with the use of ritualized dances, and will last for the life of the pair. A breeding season can take over a year from laying to fledging, with a single egg laid in each breeding attempt. Albatrosses are highly efficient in the air, using dynamic soaring and slope soaring to cover great distances with little exertion. They feed on squid, fish, and krill by either scavenging surface seizing, or diving.

The albatrosses are usually regarded as falling into four genera, but there is disagreement over the number of species. The four genera are the great albatrosses (Diomedea), the mollymawks (Thalassarche), the North Pacific albatrosses (Phoebastria), and the sooty albatrosses or sooties (Phoebetria).

Of the 21 species of albatrosses recognized by the IUCN, 19 are threatened with extinction. Numbers of albatrosses have declined in the past due to harvesting for feathers, but today the albatrosses are threatened by introduced species such as rats and feral cats that attack eggs, chicks and nesting adults; by pollution; by a serious decline in fish stocks in many regions largely due to overfishing; and by long-line fishing. Long-line fisheries pose the greatest threat, as feeding birds are attracted to the bait and become hooked on the lines and drown. Governments, conservation organizations and fishermen are all working towards reducing this by-catch.

Albatross Biology

Morphology and Flight

Unlike most Procellariiformes, albatrosses, like this Black-footed Albatross, can walk well on land

The albatrosses are a group of large to very large birds; they are the largest of the procellariiformes. The bill (beak) is large, strong and sharp-edged, the upper mandible terminating in a large hook. This bill is composed of several horny plates, and along the sides are the two "tubes," long nostrils that give the order its name. These tubes allow the albatrosses to have an acute sense of smell, an unusual ability for birds. Like other Procellariiformes, they use this olfactory ability while foraging in order to locate potential food sources. The tubes of all albatrosses are along the sides of the bill, unlike the rest of the Procellariiformes where the tubes run along the top of the bill.

The feet have no hind toe and the three anterior toes are completely webbed. The legs are strong for Procellariiformes, in fact, almost unique among the order in that they and the giant petrels are able to walk well on land.

The adult plumage of most of the albatrosses is usually some variation of dark upper-wing and back, white undersides, often compared to that of a gull. Of these, the species range from the Southern Royal Albatross which is almost completely white except for the ends and trailing edges of the wings in fully mature males, to the Amsterdam Albatross which has an almost juvenile-like breeding plumage with a great deal of brown, particularly a strong brown band around the chest. Several species of mollymawks and North Pacific albatrosses have face markings like eye patches, or have gray or yellow on the head and nape. Three albatross species, the Black-footed Albatross and the two sooty albatrosses, vary completely from the usual patterns and are almost entirely dark brown (or dark gray in places in the case of the Light-mantled Sooty Albatross). Albatrosses take several years to get their full adult breeding plumage.

The wingspans of the largest great albatrosses (genus *Diomedea*) are the greatest of any bird, exceeding 340 cm (over 11 feet), although the other species' wingspans are considerably smaller. The wings are stiff and cambered, with thickened streamlined leading edges.

Albatrosses travel huge distances with two techniques used by many long-winged seabirds, dynamic soaring and slope soaring. *Dynamic soaring* enables them to minimize the effort needed by gliding across wave fronts gaining energy from the vertical wind gradient. *Slope soaring* is more straightforward: the albatross turns to the wind, gaining height, from where it can then glide back down to the sea. Albatross have high glide ratios, around 1:22 to 1:23, meaning that for every meter they drop, they can travel forward 22 meters. They are aided in soaring by a shoulder-lock, a sheet of tendon that locks the wing when fully extended, allowing the wing to be kept up and out without any muscle expenditure, a morphological adaptation they share with the giant petrels (Pennycuick 1982). Albatrosses combine these soaring techniques with the use of predictable weather systems; albatrosses in the southern hemisphere flying north from their colonies will take a clockwise route, and those flying south will fly counter clockwise.

Taking off is one of the main times albatrosses use flapping in order to fly, and is the most energetically demanding part of a journey

Albatrosses are so well adapted to this lifestyle that their heart rates while flying are close to their basal heart rate when resting. This efficiency is such that the most energetically demanding aspect of a foraging trip is not the distance covered, but the landings, take-offs and hunting they undertake having found a food source. This efficient, long-distance traveling underlies the albatross's success as a long-distance forager, covering great distances and expending little energy looking for patchily distributed food sources.

Their adaptation to gliding flight makes them dependent on wind and waves, however, as their long wings are ill-suited to powered flight and most species lack the muscles and energy to undertake sustained flapping flight. Albatrosses in calm seas are forced to rest on the ocean's surface until the wind picks up again. They also sleep while resting on the surface (and not while on the wing as is sometimes thought). The North Pacific albatrosses can use a flight style known as flap-gliding, where the bird progresses by bursts of flapping followed by gliding. When taking off, albatrosses need to take a run up to allow enough air to move under the wing to provide lift.

Distribution and Range at Sea

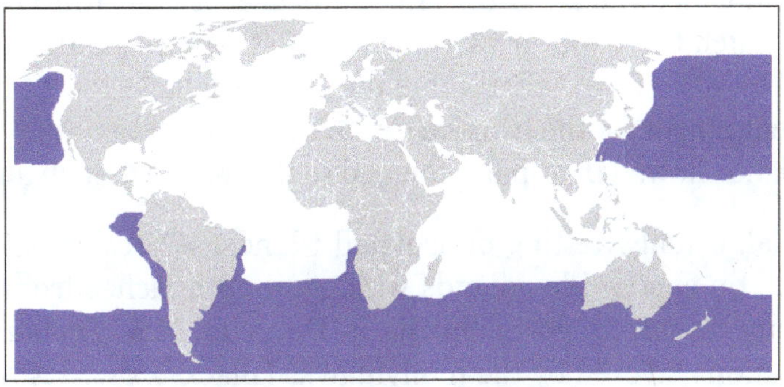

The distribution of albatrosses across the world

Most albatrosses range in the southern hemisphere from Antarctica to Australia, South Africa, and South America. The exceptions to this are the four North Pacific albatrosses, of which three occur exclusively in the North Pacific, from Hawaii to Japan, Alaska, and California; and one, the Waved Albatross, breeds in the Galapagos Islands and feeds off the coast of South America. The need for wind in order to glide is the reason albatrosses are for the most part confined to higher latitudes; being unsuited to sustain flapping flight makes crossing the doldrums extremely difficult. The exception, the Waved Albatross, is able to live in the equatorial waters around the Galapagos Islands because of the cool waters of the Humboldt Current and the resulting winds.

Albatrosses range over huge areas of ocean and regularly circle the globe

It is not known for certain why the albatrosses became extinct in the North Atlantic, although rising sea levels due to an interglacial warming period are thought to have submerged the site of a Short-tailed Albatross colony that has been excavated in Bermuda. Some southern species have occasionally turned up as vagrants in the North Atlantic and can become exiled, remaining there for decades. One of these exiles, a Black-browed Albatross, returned to Northern Gannet (a sea-bird) colonies in Scotland for many years in a lonely attempt to breed.

The use of satellite tracking is teaching scientists a great deal about the way albatrosses forage across the ocean in order to find food. They undertake no annual migration, but disperse widely after breeding, in the case of southern hemisphere species, often undertaking circumpolar trips. There is also evidence that there is separation of the ranges of different species at sea. A comparison of the foraging niches of two related species that breed on Campbell Island, the Campbell Albatross and the Grey-headed Albatross, showed the Campbell Albatross primarily fed over the Campbell Plateau whereas the Grey-Headed Albatross fed in more pelagic, oceanic waters. Wandering Albatrosses also react strongly to bathymetry, feeding only in waters deeper than 1000 m (3281 feet); so rigidly did the satellite plots match this contour that one scientist remarked, and "It almost appears as if the birds notice and obey a 'No Entry' sign where the water shallows to less than 1000 m). There is also evidence of different ranges for the two sexes of the same species; a study of Tristan Albatrosses breeding on Gough Island showed that males foraged to the west of Gough and females to the east.

Albatrosses are colonial, usually nesting on isolated islands. Where colonies are on larger landmasses, they are found on exposed headlands with good approaches from the sea in several directions, like the colony on the Otago Peninsula in Dunedin, New Zealand. Colonies vary from the very dense aggregations favored by the mollymawks (Black-browed Albatross colonies on the Falkland Islands have densities of 70 nests per 100 m^2) to the much looser groups and widely spaced individual nests favored by the sooty and great albatrosses. All albatross colonies are on islands that historically were free of land mammals.

Diet

The albatross diet is dominated by cephalopods, fish, and crustaceans (such as krill), although they will also scavenge carrion (carcasses of a dead animal) and feed on other zooplankton beyond krill. It should be noted that for most species, a comprehensive understanding of diet is only known for the breeding season, when the albatrosses regularly return to land and study is possible. The importance of each of these food sources varies from species to species, and even from population to population; some concentrate on squid alone, others take more krill or fish. Of the two albatross species found in Hawaii, one, the Black-footed Albatross, takes mostly fish while the Laysan feeds on squid.

The uses of data loggers at sea that record ingestion of water against time (providing a likely time of feeding) suggest that albatross predominantly feed during the day. Analysis of the squid beaks regurgitated by albatrosses has shown that many of the squid eaten are too large to have been caught alive and include mid-water species likely to be beyond the reach of albatross, suggesting that, for some species (like the Wandering Albatross), scavenged squid may be an important part of the diet. The source of these dead squid is a matter of debate; some certainly comes from squid fisheries, but in nature it primarily comes from the die-off that occurs after squid spawning and the vomit of squid-eating whales (sperm whales, pilot whales and Southern Bottlenose Whales). The diet of other species, like the Black-browed Albatross or the Grey-headed Albatross, is rich with smaller species of squid that tend to sink after death, and scavenging is not assumed to play a large role in their diet.

Light-mantled Sooty Albatrosses regularly dive in order to feed and can dive to below 12 m

Until recently, it was thought that albatross were predominantly surface feeders, swimming at the surface and snapping up squid and fish pushed to the surface by currents, predators, or death. The deployment of capillary depth recorders, which record the maximum dive depth undertaken by a bird (between attaching it to a bird and recovering it when it returns to land), has shown that while some species, like the Wandering Albatross, do not dive deeper than a meter, some species, like the Light-mantled Sooty Albatross, have a mean diving depth of almost 5 m and can dive as deep as 12.5 m. In addition to surface feeding and diving, they have now also been observed plunge diving from the air to snatch prey.

Breeding

Albatrosses are highly philopatric, meaning they will usually return to their natal colony to breed. This tendency to return is so strong that a study of Laysan Albatross showed that the average distance between hatching site and the site where a bird established its own territory was 22 meters.

Like most seabirds, albatrosses are K-selected (rather than R-selected) with regard to their life history, meaning they live much longer than other birds, they delay breeding for longer, and invest more effort into fewer young. Albatrosses are very long lived; most species survive upwards of 50 years, the oldest recorded being a Northern Royal Albatross that was ringed as an adult and survived for another 51 years, giving it an estimated age of 61. Given that most albatross ringing projects are considerably younger than that, it is thought likely that other species will prove to live that long and even longer.

Sky-pointing is one of the stereotyped actions of Laysan Albatross breeding dances

Albatrosses reach sexual maturity after about five years, but even once they have reached maturity, they will not begin to breed for another couple of years (even up to ten years for some species). Young non-breeders will attend a colony prior to beginning to breed, spending many years practicing the elaborate breeding rituals and "dances" for which the family is famous. Birds arriving back at the colony for the first time already have the stereotyped behaviors that compose albatross language, but can neither "read" that behavior as exhibited by other birds nor respond appropriately. After a period of trial and error learning, the young birds learn the syntax and perfect the dances. This language is mastered more rapidly if the younger birds are around older birds.

The repertoire of mating behavior involves synchronized performances of various actions such as preening, pointing, calling, bill clacking, staring, and combinations of such behaviors (like the sky-call). When a bird first returns to the colony, it will dance with many partners, but after a number of years the number of birds an individual will interact with drops, until one partner is chosen and a pair is formed. They then continue to perfect an individual language that will eventually be unique to that one pair. Having established a pair bond that will last for life, however, most of that dance will never be used ever again.

Albatrosses are thought to undertake these elaborate and painstaking rituals to ensure that the correct partner has been chosen and to perfect recognition of their partner, as egg laying and chick rearing is a huge investment. Even species that can complete an egg-laying cycle in under a year seldom lay eggs in consecutive years. The great albatrosses (like the Wandering Albatross) take over a year to raise a chick from laying to fledging. Albatrosses lay a single egg in a breeding season; if the egg is lost to predators or accidentally broken, then no further breeding attempts are made that year. The "divorce" of a pair is a rare occurrence, usually only happening after several years of breeding failure.

All the southern albatrosses create large nests for their egg, whereas the three species in the north Pacific make more rudimentary nests. The Waved Albatross, on the other hand, makes no nest and will even move its egg around the pair's territory, as much as 50 m, sometimes causing it to lose the egg. In all albatross species, both parents incubate the egg in stints that last between one day and three weeks. Incubation lasts around 70 to 80 days (longer for the larger albatrosses), the longest incubation period of any bird. It can be an energetically demanding process, with the adult losing as much as 83 g of body weight a day.

After hatching, the chick is brooded and guarded for three weeks until it is large enough to defend and thermoregulate itself. During this period, the parents feed the chick small meals when they relieve each other from duty. After the brooding period is over, the chick is fed in regular intervals by both parents. The parents adopt alternative patterns of short and long foraging trips, providing meals that weigh around 12 percent of their body weight (around 600 g). The meals are composed of fresh squid, fish, and krill, as well as stomach oil, an energy-rich food that is lighter to carry than undigested prey items. This oil is created in a stomach organ known as a proventriculus from digested prey items by most tubenoses, and gives them their distinctive musty smell.

Albatross chicks take a long time to fledge. In the case of the great albatrosses, it can take up to 280 days; even for the smaller albatrosses, it takes anywhere between 140 and 170 days. Like many seabirds, albatross chicks will gain enough weight to be heavier than their parents, and prior to fledging they use these reserves to build up body condition (particularly growing all their flight feathers), usually fledging at the same weight as their parents. Albatross chicks fledge on their own and receive no further help from their parents, who return to the nest after fledging, unaware their chick has left. Studies of juveniles dispersing at sea have suggested an innate migration behavior, a genetically coded navigation route, which helps young birds when they are first out at sea.

Albatrosses and Humans

Albatrosses and Culture

Albatrosses have been described as "the most legendary of all birds". An albatross is a central emblem in The Rime of the Ancient Mariner by Samuel Taylor Coleridge; a captive albatross is also a metaphor for the poète maudit in a poem of Charles Baudelaire. It is from the former poem that the usage of albatross as a metaphor is derived; someone with a burden or obstacle is said to have 'an albatross around their neck', the punishment given in the poem to the mariner who killed the albatross. In part due to the poem, there is a widespread myth that sailors believe it disastrous to shoot or harm an albatross; in truth, however, sailors regularly killed and ate them, but they were often regarded as the souls of lost sailors.

Albatrosses are popular birds for birdwatchers and their colonies popular destinations for ecotourists. Regular birdwatching trips are taken out of many coastal towns and cities, like Monterey and Wollongong in New South Wales, Kaikoura in New Zealand, and Sydney in Australia, where pelagic seabirds and albatrosses are easily attracted to these sightseeing boats by the deployment of fish oil into the sea. Visits to colonies can be very popular; the Northern Royal Albatross colony at Taiaroa Head in New Zealand attracts 40,000 visitors a year, and more isolated colonies are regular attractions on cruises to sub-Antarctic islands.

Threats and Conservation

In spite of often being accorded legendary status, albatrosses have not escaped either indirect or direct pressure from humans. Early encounters with albatrosses by Polynesians and Aleut Indians resulted in hunting and in some cases extirpation from some islands (such as Easter Island). As Europeans began sailing the world, they too began to hunt albatross, "fishing" for them from boats to serve at the table or blasting them for sport. This sport reached its peak on emigration lines bound for Australia, and only died down when ships became too fast to fish from, and regulations stopped the discharge of weapons for safety reasons. In the nineteenth century, albatross colonies, particularly those in the North Pacific, were harvested for the feather trade, leading to the near extinction of the Short-tailed Albatross.

This Black-browed Albatross has been hooked on a long-line

The 21 albatross species recognised by the World Conservation Union (IUCN) on their IUCN Red List, 19 are threatened, and the other two are *near threatened.* Two species (as recognized by the IUCN) are considered critically endangered: the Amsterdam Albatross and the Chatham Albatross. One of the main threats is commercial long-line fishing, as the albatrosses and other seabirds, which will readily feed on offal (internal organs used as bait), are attracted to the set bait, become hooked on the lines, and drown. An estimated 100,000 albatross per year are killed in this fashion. Unregulated pirate (illegal) fisheries exacerbate the problem.

Another threat to albatrosses is introduced species, such as rats or feral cats, which directly attack the albatross or its chicks and eggs. Albatrosses have evolved to breed on islands where land mammals are absent and have not evolved defences against them. Even species as small as mice can be detrimental; on Gough Island the chicks of Tristan Albatrosses are attacked and eaten alive by introduced house mice that are almost 300 times smaller than they are. Introduced species can have other indirect effects: cattle overgrazed essential cover on Amsterdam Island threatening the Amsterdam Albatross; on other islands introduced plants reduce potential nesting habitat.

The remains of this Laysan Albatross chick show the plastic ingested prior to death, including a bottle cap and lighter

Ingestion of plastic flotsam is another problem, one faced by many seabirds. The amount of plastic in the seas has increased dramatically since the first record in the 1960s, coming from waste discarded by ships, offshore dumping, litter on beaches, and waste washed to sea by rivers. Such plastic is impossible to digest and takes up space in the stomach or gizzard that should be used for food, or can cause an obstruction that starves the bird directly. Studies of birds in the North Pacific have shown that ingestion of plastics results in declining body weight and body condition. This plastic is sometimes regurgitated and fed to chicks; a study of Laysan Albatross chicks on Midway Atoll showed large amounts of ingested plastic in naturally dead chicks compared to healthy chicks killed in accidents. While not the direct cause of death, this plastic causes physiological stress and likely causes the chick to feel full during feedings, reducing its food intake and the chances of survival.

Scientists and conservationists (particularly BirdLife International and their partners, who run the Save the Albatross campaign) are working with governments and fishermen to find solutions to the threats albatrosses face. Techniques such as setting long-line bait at night, dying the bait blue, setting the bait underwater, increasing the amount of weight on lines and using bird scarers can all reduce the seabird by-catch For example, a collaborative study between scientists and fishermen in New Zealand successfully tested an underwater setting device for long-liners that set the lines below the reach of vulnerable albatross species. The use of some of these techniques in the Patagonian Toothfish fishery in the Falkland Islands is thought to have reduced the number of Black-browed Albatross taken by the fleet in the last 10 years.

One important step towards protecting albatrosses and other seabirds is the 2001 treaty the Agreement on the Conservation of Albatrosses and Petrels, which came into force in 2004 and has been ratified by eight countries, Australia, Ecuador, New Zealand, Spain, South Africa, France, Peru and the United Kingdom. The treaty requires these countries to take specific actions to reduce by-catch, pollution, and to remove introduced species from nesting islands. The treaty has also been signed but not ratified by another three countries, Argentina, Brazil, and Chile.

Conservationists have also worked on the field of island restoration, removing introduced species that threaten native wildlife, which protects albatrosses from introduced predators.

Species

The World Conservation Union (IUCN) and BirdLife International, among others, recognize the interim taxonomy of 21 extant species. These are the following, arranged in the four recognized genera:

- Great albatrosses *(Diomedea):*
    - Wandering Albatross *D. exulans*
    - Antipodean Albatross *D. (exulans) antipodensis*
    - Amsterdam Albatross *D. (exulans) amsterdamensis*
    - Tristan Albatross *D. (exulans) dabbenena*
    - Northern Royal Albatross *D. (epomorpha) sanfordi*
    - Southern Royal Albatross *D. epomophora*
- North Pacific albatrosses *(Phoebastria):*
    - Waved Albatross *P. irrorata*
    - Short-tailed Albatross *P. albatrus*
    - Black-footed Albatross *P. nigripes*
    - Laysan Albatross *P. immutabilis*
- Mollymawks *(Thalassarche):*
    - Black-browed Albatross *T. melanophris*
    - Campbell Albatross *T. (melanophris) impavida*
    - Shy Albatross *T. cauta*
    - Chatham Albatross *T. (cauta) eremita*

   - Salvin's Albatross *T. (cauta) salvini*
   - Grey-headed Albatross *T. chrysostoma*
   - Atlantic Yellow-nosed Albatross *T. chlororhynchos*
   - Indian Yellow-nosed Albatross *T. (chlororhynchos) carteri*
   - Buller's Albatross *T. bulleri*
- Sooty albatrosses *(Phoebetria):*
   - Dark-mantled Sooty Albatross *P. fusca*
   - Light-mantled Sooty Albatross *P. palpebrata*

Mollymawk

They are any of various oceanic birds, as the fulmar or albatross. The mollymawks are a group of medium sized albatrosses that form the genus Thalassarche. The name has sometimes been used for the genus Phoebetria as well, but these are correctly called sooty albatrosses. They are restricted to the Southern Hemisphere, where they are the most common of the albatrosses. They were long considered to be in the same genus as the great albatrosses, Diomedea, but a study of their mitochondrial DNA showed that they are a monophyletic taxon related to the sooty albatrosses, and they were placed in their own genus.

The mollymawks are a group of medium sized albatrosses that form the genus *Thalassarche*. The name has sometimes been used for the genus *Phoebetria* as well, but these are correctly called sooty albatrosses. They are restricted to the Southern Hemisphere, where they are the most common of the albatrosses.

They were long considered to be in the same genus as the great albatrosses, *Diomedea*, but a study of their mitochondrial DNA showed that they are a monophyletic taxon related to the sooty albatrosses, and they were placed in their own genus.

Species

- *Thalassarche chlororhynchos*, Atlantic Yellow-nosed Albatross
- *Thalassarche carteri*, Indian Yellow-nosed Albatross
- *Thalassarche bulleri*, Buller's Albatross
- *Thalassarche cauta*, Shy Albatross
- *Thalassarche steadi*, White-capped Albatross
- *Thalassarche eremita*, Chatham Albatross
- *Thalassarche salvini*, Salvin's Albatross
- *Thalassarche impavida*, Campbell Albatross
- *Thalassarche chrysostoma*, Grey-headed Albatross
- *Thalassarche melanophrys*, Black-browed Albatross
- *Thalassarche thyridata*

Mollymawks have the largest range in size of all the Albatross Genera, as their wingspans are 180–256 cm (71–100 in). Mollymawks have what has been described as gull-like plumage, with dark black backs, mantle and tails and lighter heads, underwings and bellies. The heads of several species are often slightly darker grey, or have dark around the eyes. They all have a colorful pinkish flesh stripe from their gape to their ear that is shown during displays. They have distinctive bill structure and coloring which makes for easier identifying than other Albatross. The bills of mollymawks are either brightly colored orange or yellow, or dark with several bright yellow lines. Mollymawk Island near South Georgia gains its name from them.

Penguin

Penguin is the common name for any of the aquatic, gregarious, flightless birds comprising the family Spheniscidae, living almost exclusively in the Southern Hemisphere, and characterized by a stout body, wings in the form of flippers, webbed feet on short legs that are placed far posterior on the body, and counter shaded dark and white plumage. While highly adapted for life in the water, being excellent swimmers and divers, penguins also spend time on land, where they breed. There are about 17 extant species of penguins, placed in six genera in the subfamily Spheniscinae.

All penguin species are native to the Southern Hemisphere, principally in cool regions, and a number breed in Antarctica and Sub-Antarctic islands. However, there also are several species that are found in the temperate zone, and one species, the Galápagos penguin, lives near the equator. Penguins are found off the coasts of South America, Africa, Australia, and New Zealand. There are no penguins based at the North Pole, as sometimes depicted in popular media; the only wild penguins in the northern hemisphere are the small group on the northernmost of the Galápagos.

Penguins provide important ecological functions are part of food chains. Most penguins feed on krill, fish, cuttlefish, squid, and other forms of sealife caught while swimming underwater. Key predators include the leopard seal, killer whale, shark, rat, and stoat.

For humans, the unique forms and behaviors of penguins have made them popular around the world. Their unusually upright, waddling gait adds to their appeal, and their striking black-and-white plumage is often likened to a tuxedo suit. Their monogamous pairing for breeding and the sacrificial parental care shown—such as the emperor penguin's walking a long distance on ice to the breeding site, undergoing substantial fasting (about 115 days for the male), and incubating the egg on their feet—adds a sympathetic quality to the birds and have led to their common depictions in various media (books, films, cartoons, television shows).

Historically, penguins have been hunted for the oil from their substantial, insulating fat layer. Aiding this has been their large breeding aggregations and the fact many penguins seem to have no special fear of humans and have approached groups of explorers without hesitation. This is likely a function of there being no large land predators in Antarctica or the nearby offshore islands that prey on or attack penguins. Today, commercial fishing, destruction of breeding habitat, collection of eggs, and predation by introduced mammals, among other anthropogenic factors have posed major threats to wild populations, of which 12 appear on the IUCN Red List of Threatened Species.

Penguin Biology

Anatomy

Orcas swim by an iceberg with Adélie penguins in the Ross Sea, Antarctica.
The Drygalski ice tongue is visible in the background

Penguins have stout bodies, with a characteristic upright posture and wings that are stiff, modified into flippers, and not folded against the body. The legs are short, placed far posterior on the body, and with webbed feet that have four forward pointing toes. Their bills are long, with impervious nares, and often laterally compressed. The feathers are small. They have 15 cervical vertebrae.

The largest living species is the emperor penguin *(Aptenodytes forsteri)*: adults average about 1.1 meters (3 feet 7 inches) tall and weigh 35 kilograms (75 pounds) or more. The smallest penguin species is the little blue penguin (also known as the fairy penguin), which stands around 40 centimeters tall (16 inches) and weighs 1 kilogram (2.2 pounds). Among extant penguins larger penguins inhabit colder regions, while smaller penguins are generally found in temperate or even tropical climates. Some prehistoric species attained enormous sizes, becoming as tall or as heavy as an adult human. These were not restricted to Antarctic regions; on the contrary, Sub-Antarctic regions harbored high diversity, and at least one giant penguin occurred in a region not quite 2,000 kilometers south of the Equator 35 million years ago, in a climate decidedly warmer than today.

All penguins are counter shaded for camouflage—that is, they have a white underside and a dark (mostly black) upper side. A predator looking up from below (such as an orca or a leopard seal) has difficulty distinguishing between a white penguin belly and the reflective water surface. The dark plumage on their backs camouflages them from aquatic predators from above.

The penguin is an accomplished swimmer, having flippers instead of wings

Penguins are superbly adapted to an aquatic life. Their wings, in the form of flippers, are useless for flight in the air but allow penguins in water to be astonishingly agile. Within the smooth plumage, a layer of air is preserved, ensuring buoyancy. The air layer also helps insulate the birds in cold waters. Penguins can drink salt water because their supraorbital gland filters excess salt from the bloodstream.

The salt is excreted in a concentrated fluid from the nasal passages. Their eyes are adapted for underwater vision, and are their primary means of locating prey and avoiding predators. They are sensitive to the various colors of the sea (violets, blues, greens) and have a second transparent eyelid that acts like goggles underwater (Saint Louis Zoo). In air it has been suggested that they are nearsighted, although research has not supported this hypothesis.

Penguins have a thick layer of insulating feathers that serve to keep them warm in water (heat loss in water is much greater than in air). The Emperor penguin, endemic to Antarctica, has the largest body mass of all penguins, which further reduces relative surface area and heat loss. They also are able to control blood flow to their extremities, reducing the amount of blood that gets cold, but still keeping the extremities from freezing. In the extreme cold of the Antarctic winter, the females are at sea fishing for food, leaving the males to brave the weather by themselves. They often huddle together to keep warm and rotate positions to make sure that each penguin gets a turn in the center of the heat pack.

Diving penguins reach 6 to 12 kilometers per hour (3.7-7.5 mph), though there are reports of velocities of 27 kilometers per hour (17 mph) (which are more realistic in the case of startled flight). The small penguins do not usually dive deep; they catch their prey near the surface in dives that normally last only one or two minutes. Larger penguins can dive deep in case of need. Dives of the large emperor penguin has been recorded that reach a depth of 565 meters (1,870 feet) and last up to 22 minutes.

On land, penguins use their tails and wings to maintain balance for their upright stance. Penguins either waddle on their feet or slide on their bellies across the snow, a movement called "tobogganing," which conserves energy while allowing them to move quickly. They also jump with both feet together if they want to move more quickly or cross steep or rocky terrain.

Penguins have an average sense of hearing for birds; this is used by parents and chicks to locate one another in crowded colonies.

Breeding

Chinstrap penguins in Antarctica

Penguins of the Antarctic and Sub-Antarctic breed on coasts and pack ice, while those closer to the equator tend to breed in coastal or forest habitats. Most species breed in large colonies, with the Macaroni penguin colony in South Georgia estimated to attract five million pairs.

Penguins form monogamous pairs for a breeding season, though the rate the same pair recouples varies drastically according to species. Nests range from holes dug in soft soil, depressions in rocks, shallow dishes of vegetation and pebbles, to the space between the top of the feet and the abdominal skin. Most penguins lay two eggs in a clutch, though the two largest species, the emperor and the king penguins, lay only one. With the exception of the emperor penguin, all penguins share the incubation duties. These incubation shifts can last days and even weeks as one member of the pair feeds at sea.

Penguin eggs are smaller than any other bird species when compared proportionally to the weight of the parent birds; at 52 grams, the little penguin egg is 4.7 percent of its mothers' weight, and the 450-gram emperor penguin egg is 2.3 percent. The relatively thick shell forms between 10 and 16 percent of the weight of a penguin egg, presumably to minimize risk of breakage in an adverse nesting environment. The yolk, too, is large, and comprises 22 to 31 percent of the egg. Some yolk often remains when a chick is born, and is thought to help sustain it if parents are delayed in returning with food.

In those penguins that lay two eggs, often the younger of two chicks will not survive. When mothers lose a chick, they sometimes attempt to "steal" another mother's chick, usually unsuccessfully as other females in the vicinity assist the defending mother in keeping her chick. In some species, such as emperor penguins, young penguins assemble in large groups called crèches.

Isabelline Adélie penguin on Gourdin Island

Systematics and Evolution

Extant penguins comprise the subfamily Spheniscinae in the family Spheniscidae. Spheniscidae traditionally was placed in the order Sphenisciformes, as the sole extant family. However, some taxonomists have enlarged the Ciconiiformes to include the Spheniscidae family, as in the Sibley-Ahlquist taxonomy. The radical and influential Sibley-Ahlquist taxonomy, based on Sibley and Ahlquist's landmark DNA-DNA hybridization studies, greatly enlarged the Ciconiiformes, adding many more families, including most of those usually regarded as belonging to the Sphenisciformes (penguins), Gaviiformes (divers), Podicipediformes (grebes), Procellariiformes (tubenosed seabirds), Charadriiformes, (waders, gulls, terns and auks), Pelecaniformes (pelicans, cormorants, gannets and allies), and the Falconiformes (diurnal birds of prey). However, many have rejected this taxonomy, and penguins commonly are listed as members of the order Sphenisciformes.

Some recent sources apply the phylogenetic taxon Spheniscidae to what here is referred to as Spheniscinae. Furthermore, they restrict the phylogenetic taxon *Sphenisciformes* to flightless taxa, and establish the phylogenetic taxon Pansphenisciformes as equivalent to the Linnean taxon Sphenisciformes, in other words, including any flying basal "proto-penguins" to be discovered eventually. Given that neither the relationships of the penguin subfamilies to each other nor the placement of the penguins in the avian phylogeny is presently resolved, the established Linnean system is thus followed here.

Living Species and Recent Extinctions

Emperor penguins *(Aptenodytes forsteri)*, the largest living species

Adelie penguin *(Pygoscelis adeliae)* feeding young

Closeup of Southern rockhopper penguin *(Eudyptes chrysocome)*

Magellanic penguins *(Spheniscus magellanicus)* guarding nest burrow. The closed neck collar denotes this species

The number of extant penguin species is debated. Depending on which authority is followed, penguin biodiversity varies between 17 and 20 living species, all in the subfamily Spheniscinae. Some sources consider the white-flippered penguin a separate *Eudyptula* species, while others treat it as a subspecies of the little penguin; the actual situation seems to be more complicated. Similarly, it is still unclear whether the royal penguin is merely a color morph of the Macaroni penguin. Also eligible to be a separate species is the Northern population of Rockhopper penguins.

Subfamily Spheniscinae—Modern Penguins

- *Aptenodytes*—Great penguins
    - King penguin, *Aptenodytes patagonicus*
    - Emperor penguin, *Aptenodytes forsteri*
- *Pygoscelis*—Brush-tailed penguins
    - Adelie penguin, *Pygoscelis adeliae*
    - Chinstrap penguin, *Pygoscelis antarctica*
    - Gentoo penguin, *Pygoscelis papua*
- *Eudyptula*—Little penguins
    - Little blue penguin, *Eudyptula minor*

- Northern little penguin, *Eudyptula albosignata* (provisional)
- *Spheniscus*—Banded penguins
  - Magellanic penguin, *Spheniscus magellanicus*
  - Humboldt penguin, *Spheniscus humboldti*
  - Galapagos penguin, *Spheniscus mendiculus*
  - African penguin, *Spheniscus demersus*
- *Megadyptes*
  - Yellow-eyed penguin, *Megadyptes antipodes*
  - Waitaha penguin, *Megadyptes waitaha* (extinct)
- *Eudyptes*—Crested penguins
  - Fiordland penguin, *Eudyptes pachyrynchus*
  - Snares penguin, *Eudyptes robustus*
  - Erect-crested penguin, *Eudyptes sclateri*
  - Southern rockhopper penguin, *Eudyptes chrysocome*
  - Northern rockhopper penguin, *Eudyptes moseleyi*
  - Royal penguin, *Eudyptes schlegeli* (disputed)
  - Macaroni penguin, *Eudyptes chrysolophus*
  - Chatham Islands penguin, *Eudyptes* sp. (extinct)

Fossil Genera - Order Sphenisciformes

- Basal and Unresolved Taxa (All Fossil)
  - *Waimanu*—basal (Middle-Late Paleocene)
  - *Perudyptes* (Middle Eocene of Atacama Desert, Peru)—basal?
  - Spheniscidae gen. sp. indet. CADIC P 21 (Leticia Middle Eocene of Punta Torcida, Argentina)
  - *Delphinornis* (Middle/Late Eocene? - Early Oligocene of Seymour Island, Antarctica) — Palaeeudyptinae, basal, new subfamily 1?
  - *Archaeospheniscus* (Middle/Late Eocene – Late Oligocene) – Palaeeudyptinae? New subfamily 2?
  - *Marambiornis* (Late Eocene –? Early Oligocene of Seymour Island, Antarctica) – Palaeeudyptinae, basal, new subfamily 1?
  - *Mesetaornis* (Late Eocene –? Early Oligocene of Seymour Island, Antarctica) – Palaeeudyptinae, basal, new subfamily 1?

- *Tonniornis* (Late Eocene –? Early Oligocene of Seymour Island, Antarctica)
- *Wimanornis* (Late Eocene –? Early Oligocene of Seymour Island, Antarctica)
- *Duntroonornis* (Late Oligocene of Otago, New Zealand) – possibly Spheniscinae
- *Korora* (Late Oligocene of S Canterbury, New Zealand)
- *Platydyptes* (Late Oligocene of New Zealand) – possibly not monophyletic; Palaeeudyptinae, Paraptenodytinae or new subfamily?
- *Spheniscus gen. et sp. indet (Late Oligocene/Early Miocene of Hakataramea, New Zealand)*
- *Madrynornis* (Puerto Madryn Late Miocene of Argentina) – possibly Spheniscinae
- *Pseudaptenodytes* (Late Miocene/Early Pliocene)
- *Dege* (Early Pliocene of South Africa) – possibly Spheniscinae
- *Marplesornis* (Early Pliocene) – possibly Spheniscinae
- *Nucleornis* (Early Pliocene of Duinfontain, South Africa) – possibly Spheniscinae
- *Inguza* (Late Pliocene) – probably Spheniscinae; formerly *Spheniscus predemersus*

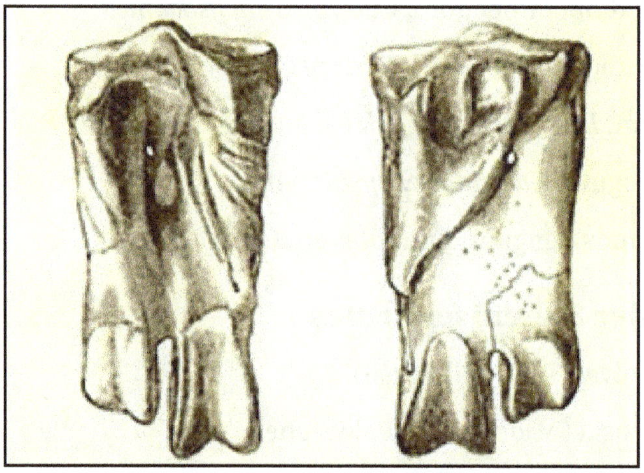

A damaged tarsometatarsus of the prehistoric Narrow-flippered Penguin *(Palaeeudyptes antarcticus)*

Family Spheniscidae

- Subfamily Palaeeudyptinae – Giant penguins (fossil)
  - *Crossvallia* (Cross Valley Late Paleocene of Seymour Island, Antarctica) – tentatively assigned to this subfamily
  - *Anthropornis* (Middle Eocene? – Early Oligocene of Seymour Island, Antarctica) – tentatively assigned to this subfamily
    - Nordenskjoeld's Giant Penguin, *Anthropornis nordenskjoeldi*
  - *Icadyptes* (Late Eocene of Atacama Desert, Peru)

   - *Palaeeudyptes* (Middle/Late Eocene – Late Oligocene) – polyphyletic; some belong in other subfamilies
   - *Pachydyptes* (Late Eocene)
   - *Anthropodyptes* (Middle Miocene) – tentatively assigned to this subfamily
- Subfamily Paraptenodytinae – Stout-footed penguins (fossil)
   - *Arthrodytes* (San Julian Late Eocene/Early Oligocene – Patagonia Early Miocene of Patagonia, Argentina)
   - *Paraptenodytes* (Early – Late Miocene/Early Pliocene)
- Subfamily Palaeospheniscinae – Slender-footed penguins (fossil)
   - *Eretiscus* (Patagonia Early Miocene of Patagonia, Argentina)
   - *Palaeospheniscus* (Early? – Late Miocene/Early Pliocene) – includes *Chubutodyptes*

The evolutionary history of penguins is extensively researched and represents a showcase of evolutionary biogeography; though as penguin bones of any one species vary much in size and few good specimens are known, the alpha taxonomy of many prehistoric forms still leaves much to be desired.

The basal penguins are considered to have lived around the time of the Cretaceous–Tertiary extinction event, somewhere in the general area of (southern) New Zealand and Byrd Land, Antarctica. Due to plate tectonics, these areas were at that time less than 1500 kilometers rather than the 4000 kilometers of today. The most recent common ancestor of penguins and their sister clade has been roughly dated to the Campanian–Maastrichtian boundary, around 70–68 million years ago. What can be said as certainly as possible in the absence of direct (that is, fossil) evidence is that by the end of the Cretaceous, the penguin lineage must have been evolutionarily well distinct, though much less so morphologically; it is fairly likely that they were not yet entirely flightless at that time, as flightless birds appear to have generally low resilience to the breakdown of trophic webs that follow the initial phase of mass extinctions because of their below-average dispersal capabilities.

Basal Fossils

The oldest known fossil penguin species is *Waimanu manneringi,* which lived in the early Paleocene epoch of New Zealand, or about 62 mya. While they were not as well-adapted to aquatic life as modern penguins, *Waimanu* were generally loon-like birds but already flightless, with short wings adapted for deep diving. They swam on the surface using mainly their feet, but the wings were—as opposed to most other diving birds, living and extinct—already adapting to underwater locomotion.

Perudyptes from northern Peru was dated to 42 mya. An unnamed fossil from Argentina indicates that by the Bartonian (Middle Eocene), some 39–38 mya, primitive penguins had spread to South America and were in the process of expanding into Atlantic waters.

Palaeëudyptines

During the Late Eocene and the Early Oligocene (40–30 mya), some lineages of gigantic penguins existed. Nordenskjoeld's giant penguin was the tallest, growing nearly 1.80 meters (6 feet) tall. The New Zealand giant penguin was probably the heaviest, weighing 80 kilograms or more. Both were found on New Zealand, the former also in the Antarctic farther eastwards.

Traditionally, most extinct species of penguins, giant or small, had been placed in the paraphyletic subfamily called Palaeeudyptinae. More recently, with new taxa being discovered and placed in the phylogeny if possible, it is becoming accepted that there were at least two major extinct lineages. One or two closely related ones occurred in Patagonia, and at least one other—which is or includes the paleeudyptines as recognized today—occurred on most Antarctic and subantarctic coasts.

But size plasticity seems to have been great at this initial stage of penguin radiation: on Seymour Island, Antarctica, for example, around 10 known species of penguins ranging in size from medium to huge apparently coexisted some 35 mya during the Priabonian (Late Eocene). It is not even known whether the gigantic palaeeudyptines constitute a monophyletic lineage, or whether gigantism was evolved independently in a much restricted Palaeeudyptinae and the Anthropornithinae—whether they were considered valid, or whether there was a wide size range present in the Palaeeudyptinae as delimited as usually done these days (that is, including *Anthropornis nordenskjoeldi*). The oldest well-described giant penguin, the 5-foot-tall *Icadyptes salasi*, actually occurred as far north as northern Peru about 36 mya.

In any case, the gigantic penguins had disappeared by the end of the Paleogene, around 25 mya. Their decline and disappearance coincided with the spread of the Squalodontoidea and other primitive, fish-eating toothed whales, which certainly competed with them for food, and were ultimately more successful. A new lineage, the Paraptenodytes, which includes smaller but decidedly stout-legged forms, had already arisen in southernmost South America by that time. The early Neogene saw the emergence of yet another morphotype in the same area, the similarly-sized but more gracile Palaeospheniscinae, as well as the radiation which gave rise to the penguin biodiversity of our time.

Origin and Systematics of Modern Penguins

Modern penguin's consititute two undisputed clades and another two more basal genera with more ambiguous relationships. The origin of the Spheniscinae lies probably in the latest Paleogene, and geographically it must have been much the same as the general area in which the order evolved: The oceans between the Australia-New Zealand region and the Antarctic. Presumedly diverging from other penguins around 40 mya, it seems that the Spheniscinae were for quite some time limited to their ancestral area, as the well-researched deposits of the Antarctic Peninsula and Patagonia have not yielded Paleogene fossils of the subfamily. Also, the earliest spheniscine lineages are those with the most southern distribution.

The genus *Aptenodytes* appears to be the basalmost divergence among living penguins; they have bright yellow-orange neck, breast, and bill patches; incubate by placing their eggs on their feet, and when they hatch the chicks are almost naked. This genus has a distribution centered on the Antarctic coasts and barely extends to some subantarctic islands today.

Pygoscelis contains species with a fairly simple black-and-white head pattern; their distribution is intermediate, centered on Antarctic coasts but extending somewhat northwards from there. In external morphology, these apparently still resemble the common ancestor of the Spheniscinae, as *Aptenodytes'* autapomorphies are in most cases fairly pronounced adaptations related to that genus' extreme habitat conditions. As the former genus, *Pygoscelis* seems to have diverged during the Bartonian, but the range expansion and radiation which led to the present-day diversity probably did not occur until much later; around the Burdigalian stage of the Early Miocene, roughly 20–15 mya.

The genera *Spheniscus* and *Eudyptula* contain species with a mostly subantarctic distribution centered on South America; some, however, range quite far northwards. They all lack carotenoid coloration, and the former genus has a conspicuous banded head pattern; they are unique among living penguins by nesting in burrows. This group probably radiated eastwards with the Antarctic Circumpolar Current out of the ancestral range of modern penguins throughout the Chattian (Late Oligocene), starting approximately 28 mya. While the two genera separated during this time, the present-day diversity is the result of a Pliocene radiation, taking place some 4–2 mya.

The *Megadyptes–Eudyptes* clade occurs at similar latitudes (though not as far north as the Galapagos Penguin), has its highest diversity in the New Zealand region, and represent a westward dispersal. They are characterized by hairy yellow ornamental head feathers; their bills are at least partly red. These two genera diverged apparently in the Middle Miocene (Langhian, roughly 15–14 mya), but again, the living species of *Eudyptes* appear the product of a later radiation, stretching from about the late Tortonian (Late Miocene, 8 mya) to the end of the Pliocene.

The geographical and temporal pattern or spheniscine evolution corresponds closely to two episodes of global cooling documented in the paleoclimatic record . The emergence of the subantarctic lineage at the end of the Bartonian corresponds with the onset of the slow period of cooling that eventually led to the ice ages some 35 million years later. With habitat on the Antarctic coasts declining, by the Priabonian more hospitable conditions for most penguins existed in the subantarctic regions rather than in Antarctica itself. Notably, the cold Antarctic Circumpolar Current also started as a continuous circumpolar flow only around 30 mya, on the one hand forcing the Antarctic cooling, and on the other facilitating the eastward expansion of *Spheniscus* to South America and eventually beyond.

Later, an interspersed period of slight warming was ended by the Middle Miocene Climate Transition, a sharp drop in global average temperature from 14–12 mya, and similar abrupt cooling events followed at 8 mya and 4 mya; by the end of the Tortonian, the Antarctic ice sheet was already much like today in volume and extent. The emergence of most of today's subantarctic penguin species appears almost certainly to have been caused by this sequence of Neogene climate shifts.

Relationship to other Bird Orders

Penguin ancestry beyond *Waimanu* remains unknown and not well-resolved by molecular or morphological analyses. The latter tend to be confounded by the strong adaptive autapomorphies of the Sphenisciformes; a sometimes perceived fairly close relationship between penguins and grebes is almost certainly an error based on both groups' strong diving adaptations, which are homoplasies. On the other hand, different DNA sequence datasets do not agree in detail with each other either.

What seems clear is that penguins belong to a clade of Neoaves (living birds except paleognaths and fowl) which comprises what is sometimes called "higher waterbirds" to distinguish them from the more ancient waterfowl. This group contains such birds as storks, rails, and the seabirds, with the possible exception of the Charadriiformes.

Inside this group, penguin relationships are far less clear. Depending on the analysis and dataset, a close relationship to Ciconiiformes or to Procellariiformes has been suggested. Some think the penguin-like plotopterids (usually considered relatives of anhingas and cormorants) may actually be a sister group of the penguins, and that penguins may have ultimately shared a common ancestor with the Pelecaniformes and consequently would have to be included in that order, or that the plotopterids were not as close to other pelecaniforms as generally assumed, which would necessitate splitting the traditional Pelecaniformes in three.

The auk of the Northern Hemisphere is superficially similar to penguins: they are not related to the penguins at all, but considered by some to be a product of moderate convergent evolution.

Shearwater

Shearwater is any member of more than a dozen species of long-winged oceanic birds belonging to the family Procellariidae (order Procellariiformes), which also includes the fulmars and the petrels. Typical shearwaters are classified in the genus Puffinus, which has approximately 20 species. Shearwaters are drab, slender-billed birds that range from 35 to 65 cm (14 to 26 inches) in length. The common name shearwater describes the birds' habit of gliding on stiff wings along the troughs of waves. The name is sometimes also applied to the skimmers, an unrelated oceanic bird family.

Shearwater

Shearwaters nest in burrows on offshore islands and coastal hills in the North Atlantic, eastern South Atlantic, and Mediterranean and throughout most of the Pacific. Colonies may number hundreds of thousands of pairs, and at night, when the calling adults move in and out of the burrows, the din is deafening. A single egg is incubated in turn by the male and female, and the chick is raised by both parents.

Several shearwater species have extremely large geographic ranges. The sooty shearwater (*P. griseus*) is about 50 cm (19.5 inches) long with a wingspread of approximately 85 cm (33 inches). It breeds near Australia, New Zealand, and southern South America and winters in the offshore waters of the Atlantic and Pacific. The common, or Manx, shearwater (*P. puffinus*), whose length is

30 to 37 cm (about 12 to 15 inches), is a large species that breeds on several islands on both sides of the North Atlantic. The slender-billed, or short-tailed, shearwater (*P. tenuirostris*), which is called *muttonbird* in Australia and *whalebird* in Alaska, breeds on islands in the Bass Strait of Australia and in Tasmania. It spends the remainder of the year circumnavigating the Pacific. The bird is about 40 cm (16 inches) long, and it is harvested in large numbers for meat and oil.

A sooty shearwater (Puffinus griseus) flying above Monterey Bay, California, U.S.

Newell's shearwater (P. newelli) is about 33 cm (13 inches) long and has a geographic range that spans a large portion of the North Pacific Ocean. The IUCN Red List of Threatened Species classified it as endangered despite the presence of several breeding colonies throughout the Hawaiian Islands. The Newell's shearwater population declined by some three-fifths after Hurricane Iniki passed over the bird's principal breeding colony on Kauai in 1992. Habitat loss caused by feral pigs and goats further contributed to the species's steady decrease in numbers.

Townshend's shearwater (P. auricularis) and the Balearic shearwater (P. mauretanicus), both also 33 cm in length, are classified as critically endangered in the IUCN Red List. Townshend's shearwater faces the greatest threat of extinction of all shearwaters, because it breeds in a single location, Socorro Island, where many individuals are preyed upon by feral cats. A population study performed in 2008 documented fewer than 100 breeding pairs on the island. Although the Balearic shearwater's summer range includes much of northwestern and southwestern Europe and the Mediterranean Sea, breeding only occurs on five of the Balearic Islands, where individuals of all ages often fall prey to cats, genets (a catlike carnivore of the family Viverridae), rats, and peregrine falcons (Falco peregrinus). In the water, Balearic shearwaters in search of food are often hooked in fishing lines as they follow commercial fishing boats.

References

- Ocean-fishes, marine-life: oceana.org, Retrieved 22 July, 2019
- Sea-turtle: newworldencyclopedia.org, Retrieved 25 January, 2019
- Marine-iguana, wildlife: galapagosconservation.org.uk, Retrieved 27 April, 2019
- Marine-iguana, reptiles, galapagos-islands: nathab.com, Retrieved 19 August, 2019
- Crocodile: newworldencyclopedia.org, Retrieved 9 March, 2019
- 1077-what-is-a-seabird: acap.aq, Retrieved 1 February, 2019

- Petrel, animal: britannica.com, Retrieved 28 June, 2019
- Snow-petrel, species: nzbirdsonline.org.nz, Retrieved 26 March, 2019
- Storm-petrel, animal: britannica.com, Retrieved 6 January, 2019
- Albatross: newworldencyclopedia.org, Retrieved 11 July, 2019
- Mollymawks: beautyofbirds.com, Retrieved 10 May, 2019
- Shearwater, animal: britannica.com, Retrieved 2 August, 2019

3

Marine Microorganisms

The marine microorganisms are broadly divided into two categories, namely, prokaryotes and eukaryotes. Marine prokaryotes are further subdivided into marine viruses, marine bacteria and marine archaea. The chapter closely examines these key categories of marine microorganisms to provide an extensive understanding of the subject.

Marine microorganisms are characterized by their territory as the microorganisms living in a marine domain, that is, in the saltwater of an ocean or sea or the bitter water of a beach front estuary. A microorganism (or organism) is any minute living creature, that is, any life shape too little for the exposed human eye to see, requiring a magnifying instrument. Microorganisms are extremely various. They can be single-celled or multicellular and incorporate all microorganisms and archaea and most protozoa, and also a few types of growths, green growth, and certain minute creatures, for example, rotifers. Numerous naturally visible creatures and plants have minuscule adolescent stages.

Marine microorganisms are the dwellers of the oceanic waters and comprises of extensive and a diverse assortment of bacteria, virus, protists and fungi. Marine microorganisms exist, not only in the surface waters of the sea, but also in the lower and abyssal depths from coastal to the offshore regions, and from the general oceanic to the specialized niches like blue waters of coral reefs to black smokers of hot thermal vents on the sea floor including oceans, estuaries and lagoons, mangroves and coral reefs, the deep sea and the sea floor. Prokaryotic organisms with bacterial r RNA and lipid membrane layers constitute the bacterial domain, while viruses having a size range of 20 – 200 nm are about 10 to 100 times smaller than most bacteria with genetic material in the form or DNA or RNA. Although microorganisms are seldom perceivable in natural habitats, their contribution amounts to half the world's biomass.

Amazing diversity of marine microbes

Unlike plant and animal diversity, microbial melange is a strenuous area of research due to the diverse physiologic lifestyle and metabolic capabilities. The study of marine microbes has emerged from a couple of decades; however the finite and dispersed details in this area demand microbial taxonomy, which is quite meticulous and effortful. Microbial diversity is not only typological but it is more incorporating and comprehensive compared to plant and animal domain. Marine micro biota exists in the sea bed floor and helps in the nutrient cycling as well as having immense potential in the production of enzymes, anti tumor compounds, and certain antibiotics. Evaluating microbial diversity is a formidable task, thus the importance and study of microbial ecology become demanding and captivating. Researches and experiments on microbial biodiversity will help in exploring different dimensions of marine microbes and futuristic use of their bioactive compounds.

Role of Marine Microorganisms

Source of Food

Low temperature, high salinity, absence of light and high pressure is characteristic features of the marine environment and heterotrophic bacteria are well adapted to them. Changes in water temperature, salinity and other physicochemical parameters are important criteria on which these bacteria depend. Heterotrophic bacteria are the major components contributing to the oceanic microbial population and also help in the biological transformation of organic matter and the generation of carbon dioxide. Thus; these bacteria not only serve as intermediates in biogeochemical processes, but also serve as an important food source for a variety of marine organisms and maintain an immaculate environment.

Utilization in the Breakdown of Organic Matter

Bacteria play a conclusive role in the breakdown of all organic matter into their original forms. Some proteolytic bacteria such eubacteria and *pseudomonas sp.* help in the breakdown of proteins. Cellulolytic bacteria, such as *Cytophaga, Sporocytophaga* are responsible for cellulose decomposition. Chitin, a structural polysaccharide is not degraded easily as it is reported to be preserved in fossils *Bacillus, Pseudomonas* and *Vibrio,* by the presence of their exoenzyme chitinase can degrade this biopolymer. *Clostridium pectinovorum* is known for the decomposition of pectin resulting in the end-products as pectic acid and methanol.

Sources of Therapeutic Enzymes

Salt-tolerant, marine organisms provide a fascinating alternative for therapeutic purposes. The diverse range of enzymatic activity and catalysing ability for various biochemical reactions and novel enzymes, are characteristic feature of marine microorganism. Halophilic microorganisms contain hydrolytic enzymes, which can lead to precipitation of most of the proteins moreover saline sea water is closer to human blood plasma, and can produce microbial products specifically enzymes which are less toxic and have low or zero side effect when therapeutic application to humans are performed.

Production of Secondary Metabolites and Bioactive Compounds

Sophisticated chemical structures are common in secondary metabolites which are produced during the idiophase of microbial growth. A unique example of secondary metabolite is antibiotic production by marine bacteria which is uncommon in terrestrial microbes. Antibiotic production by marine bacteria was first studied by Rosenfeld and Zobell and consequently many others have studied the same. An interesting area of bacteriological research is the study of antagonistic properties towards human pathogens, some of which are recently discovered and are MDR strains (Multi drug resistant) or already resistant to earlier discovered antibiotics. The sea is a rich source of useful compounds with new chemical structures including significant immunomodulation, useful against allergy, anti- inflammatory and as a consequence, antitumor and analgesic, antibacterial and antiviral activities· A large proportion of active compounds, including fucoxanthin, astaxanthin, marine collagen peptides, dieckol, Krill oil and fucoidan, have been already researched to against metabolic dysfunction such as diabetes and obesity. The effects of marine bioactive compounds when tested showed no toxicity and side effects. Marine bioactives have the potential to develop as functional food or drugs, as their biological activities appear to influence the pathogenesis and the clinical course of several metabolic diseases. Further research should be carried out in this area to disclose newer preventive and potential therapeutic strategies against metabolic dysfunction. Table, shows some bioactive compounds with their source and mechanism of action.

Different Bioactive Compounds

Table: Bioactive compounds with their source and mechanism of action.

Compound	Source	Effect/Mechanism of action
Fucoxanthin	Kelp	Antidiabetic/Anti-obesity activity
Astaxanthin	Marine algae/marine phytoplankton	Antidiabetic/Anti-obesity activity
Marine collagen peptides	Deep-sea fish	Antidiabetic activity
Dieckol	Ecklonia cava	Antidiabetic/Anti- hyperlipidemic effect
Krill oil	Antarctic krill	Antidiabetic/ Anti-obesity activity
Fucoidan	Laminaria sp.	Antidiabetic/ Anti-obesity activity
Hyrtiosal	Marine sponge	Inhibit protein tyrosine phosphatase 1B (PTP1B) activity
SQA & SHQA	Sargassum yezoense	Reduce insulin resistance

Role of Microbes in Marine Biogeochemical Cycles

Biogeochemical cycles refer to the movement and transfer of various chemical elements between several compartments comprising of living and non-living systems. Marine microbe's breakdown the complex compounds present in the marine and estuarine ecosystem. The cellulose present in the mangrove and associate vegetation (blue carbon) is broken to carbon dioxide (CO_2) by cellulolytic bacteria that are abundantly available in the intertidal mudflats. Nitrogen fixing blue green algae and denitrifying bacteria present in the marine and estuarine ecosystem play a critical role in regulating and maintaining the nitrogen balance in the atmosphere. The stability and proportions of different gases in the atmosphere are mostly regulated by marine microbes.

Application in Drugs and Pharmaceutics

It is apprehended that oceans involve a huge number of natural products and novel chemical structures with unique biological activities useful in finding the prospective drugs with pronounced efficacy and specificity for the treatment of human diseases pharmaceutical products such as novel anti-inflammatory agents (e.g., pseudopterosins, topsentins, scytonemin, and manoalide), anticancer agents (e.g., bryostatins, discodermolide, eleutherobin, and sarcodictyin), and antibiotics (e.g., marinone). The contribution of probiotic bacteria, such as *lactobacilli* and *bifidobacteria,* is mainly in the control of pathogenic microbes, through production of antibacterial protein namely, bacteriocin and anticancer substances. The dietary supplements of *lactobacilli* are reportedly decreasing the induction of experimental colon cancer. Most of the marine animal phyla produce toxins and some studies show that these marine toxins may be produced by marine bacteria associated the animals. The microbial toxins are useful in neurophysiological and neuropharmacological studies. For example, bacteria present in *Noctiluca scintillans* are responsible for causing red tides.

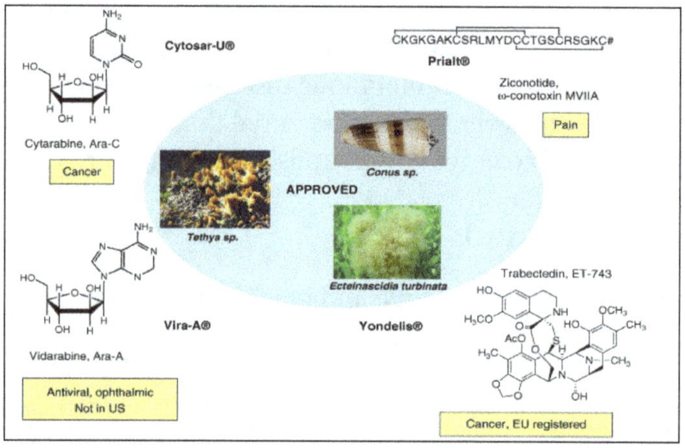

Marine natural products or derivatives thereof approved for use by the FDA or EMEA, their biological source, chemical structures and treatment usage

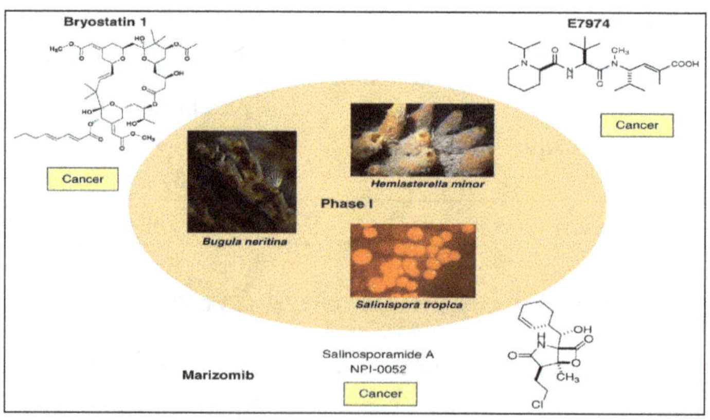

Marine natural products

Prokaryotes

Marine Viruses

Viruses are the most abundant biological entities in the oceans, with as many as 10 million in a single teaspoon of seawater. If all the viruses in the world's oceans were laid end-to-end they would span 100 times the distance across our galaxy. Not only are viruses hugely abundant, they are also major players in the world's oceans. They play a key role in global biogeochemical cycling, in which substances like carbon and phosphorous move between living organisms and the atmosphere and oceans. Viruses also represent a huge reservoir of unexplored genetic diversity and they drive the evolution of their hosts through horizontal gene transfer between individuals.

However, these roles in evolution and their other beneficial influences are often overlooked, as viruses are mostly viewed as pathogens. This leaves viruses as the least studied of the biota living within coral reefs, making them the 'dark matter' component of reef ecosystems.

Viruses can be found wherever life is found and in seawater they outnumber Bacteria, Archaea, and Eukaryote combined by an order of magnitude or more. The concentration of virions in surface seawater in the open ocean is on the order of 5–10 billion perliter. Their concentration declines with depth, butviruses are present in the deepest waters sampled (44000 m).The concentration of virions in surface marine sediments is typically even higher than in the immediate overlying waters. As in the water column, viruses become less abundant with depth in the sediment, but remain detectable to depths 4100 m below the seafloor and in sediments with estimated ages of hundreds of thousands of years or more.

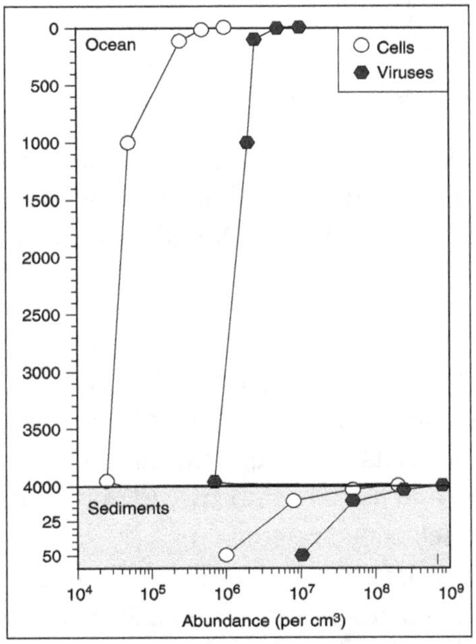

Idealized plot of the concentrations of viruses and cells in open ocean seawater and sediments. Sediments are indicated by the shaded area. Sediment depth (in bold) starts over from zero at the sediment–water interface and is plotted at a different scale for clarity. "Cells" refers to all cells, but these numbers are dominated by bacteria and archaea. The numbers shown are approximate and concentrations at a given depth can vary by an order of magnitude depending on location. Concentrations tend to be higher in areas of high productivity such as the coastal zone.

Virions suspended in seawater are a component of the marine plankton and are therefore often referred to as virioplankton. Those in sediments are sometimes referred to as viriobenthos. Plankton ecologists find it convenient at times to refer to plankton according to which of seven different size categories they belong. Because of their small size, most of the viruses in seawater are members of the smallest named size category, the femtoplankton (0.02–0.2 mm), but larger viruses are also part of the picoplankton (0.2–2 mm), a size class that also includes almost all of the bacteria and archaea and the smallest protists.

Viral infections have been reported for a broad range of marine organisms and these infections have wide-ranging effects on everything from evolution, ecology, and ocean biogeochemistry to human health and economics. Viral infections of phytoplankton, the foundation of the marine.

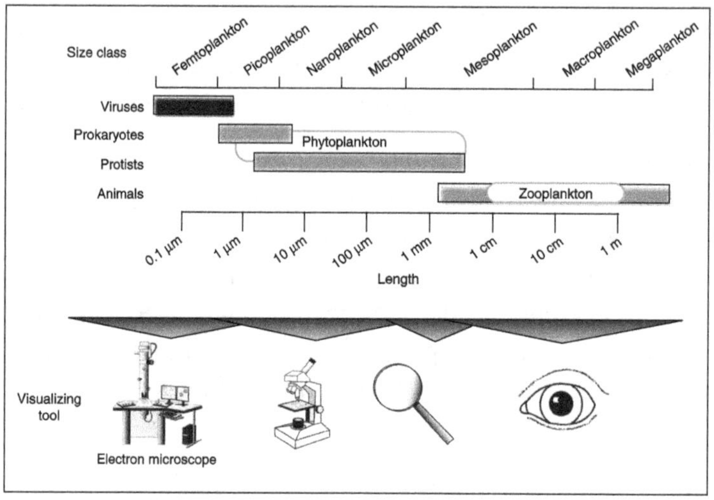

Viruses as members of the marine plankton

As a group, viruses are the smallest and most abundant members of the plankton and an electron microscope is required to reveal their details. Plankton are often named by the size category to which they belong, but may also be referred to in functional categories such as virioplankton, phytoplankton, and zooplankton.

Food web, affect how much of the ocean's primary production is transferred to higher trophic levels. Large increases in phytoplankton concentrations (blooms) are a common phenomenon in the ocean and these periodic in- creases in primary production serve as a critical food supply for life in the sea. Viruses also spread more easily through populations at high density, however, and phytoplankton blooms can be decimated by the rapid spread of viral infection. Lysis of phytoplankton decreases the efficiency with which they are consumed by grazers and much of the primary production is instead decomposed by bacteria. Bacteria are themselves susceptible to viral infections and viral lysis of bacteria likewise diverts material and energy from protistan grazers, which favors further growth and respiration by other uninfected bacteria. It thus appears that the net effect of lytic viral activity in the marine food web is to decrease the amount of carbon that reaches higher trophic levels (e.g., fish) while increasing the amount of carbon, which is respired by the bacteria.

From the perspective of humans, viruses could be lamented as reducing the efficiency of wild fisheries, but by how much is unknown. In any case, there is little to be done about the persistent, ubiquitous influence of viruses on the ocean's food web; it is an inescapable feature of the natural cycle of life in the sea. However, viruses are a legitimate human concern with tangible economic consequences when they cause death and disease among farmed marine species such as shrimp and fish. As wild fisheries become increasingly threatened by overexploitation, humans are relying more heavily on aqua- and Mari culture as a source of protein. Intensive Mari culture involves growth of single species at high densities, which, unfortunately, is a condition under which viral and other diseases can easily spread within the farm and from the farm to wild populations. The economic losses resulting from viral diseases in aquaculture are estimated to be in the billions of dollars annually so there is a strong incentive to find solutions to combat viral diseases in commercially exploited marine species.

Marine Bacteria

Marine bacteria are single-celled organisms that can be shaped like little spheres, rods, or (less commonly) spirals. They are often very small, with cell diameters of just a few microns (about 1/100th the width of a human hair). They perform all kinds of chemical processes in the open ocean, including most of the steps in nitrogen cycling.

Cyanobacteria are a large group of photosynthetic bacteria, some of which can "fix" nitrogen, converting nitrogen gas into more biologically useful compounds. Cyanobacteria live in all kinds of environments, but are especially important in open-ocean ecosystems. They were formerly known as "blue-green algae" but are now recognized as a type of bacteria, not a type of algae (algae are eukaryotes).

Marine Bacteria that "Fix" Nitrogen

Trichodesmium

Microphotograph by Angel White

Trichodesimium is a genus of colonial cyanobacteria that is one of the most important and well-studied nitrogen-fixing organisms found in open-ocean areas such as Station ALOHA. It is one of the few organisms involved in the oceanic nitrogen cycle that is visible to the naked eye. Trichodesmium is a colonial organism that forms hair-like strands, which sometimes aggregate into tiny "puffballs" up to a millimeter or two across. When winds are light, Trichodesmium colonies may clump together and float right on the sea surface, where they are known as "sea sawdust."

Trichodesmium uses an enzyme called "nitrogenase" to transform nitrogen gas into more biologically useful compounds (a process called "nitrogen fixation.") However, nitrogenase is inactivated in the presence of oxygen (which is produced by many photosynthetic organisms as a byproduct of photosynthesis). Because of this, most nitrogen-fixing microbes separate the processes of nitrogen fixation and photosynthesis either spatially (using different types of cells for each process), or temporally, by performing photosynthesis in the daytime and fixing nitrogen at night.

Trichodesmium, however, uses neither of these strategies. It is able to fix nitrogen during the daytime, but does not have specialized cells to perform the job. Researchers are very interested in figuring out how Trichodesmium is able to fix nitrogen in the daytime.

During the BioLINCS cruise, Trichodesmium is being studied using microscopy, nitrogen-uptake experiments, and nucleic acid (DNA and RNA) analyses. These analyses will be carried both out on the ship and later on shore. Among other things, researchers will be using genetic "probes" that look for a gene called nifH, which is one of several genes that help microbes produce the nitrogenase enzyme. This will help them determine the abundance and nitrogen-fixing ability of Trichodesmium. The Environmental Sample Processor (ESP) will also be using the nifH gene to monitor the presence of Trichodesmium in real time.

Heterocystus Cyanobacteria

Heterocystus cyanobacteria are multi-celled organisms that form microscopic filaments and perform nitrogen fixation in the open ocean. The most common genus of heterocystus cyanobacteria in open-ocean areas is *Richelia,* which is almost always found living inside of diatoms, a type of microscopic marine algae.

As part of this "symbiotic" living arrangement, the heterocystus cyanobacteria provide the diatoms with "fixed" nitrogen and other nutrients. It is not entirely clear what the bacteria get out of the arrangement.

Like all nitrogen fixing organisms, heterocystus cyanobacteria use an enzyme called "nitrogenase" to transform nitrogen gas into more useable forms of nitrogen. However, nitrogenase is inhibited in the oxygen-rich environment that exists inside most cells. As multi-cellular organisms, heterocystus cyanobacteria have evolved specialized cells called "heterocysts," which provide an environment more conducive to nitrogen fixation.

During the BioLINCS cruise, heterocystus cyanobacteria are being studied using DNA analyses that are carried out manually on board the ship and performed automatically within the Environmental Sample Processor (ESP). Among other things, researchers will be looking for the nifH gene in Richelia diatoms. This is one of several genes that allow cyanobacteria to produce the nitrogenase enzyme. Researchers will also be combining DNA analyses with measurements using influx flow cytometer (IFC) to determine whether heterocystus cyanobacteria are associated with specific types of marine algae.

Crocosphaera

Crocosphaera is a genus of singled-celled cyanobacteria that lives through photosynthesis and can "fix" nitrogen. It grows in many tropical ocean areas where the water is about 24° Celsius (75° Fahrenheit). At about two to four microns across, Crocosphaera is much smaller than Trichodesmium. Like all nitrogen-fixing organisms, Crocosphaera uses an enzyme called "nitrogenase" to transform nitrogen gas into more biologically useful compounds. However, nitrogenase doesn't work well in the presence of oxygen (a byproduct of photosynthesis). As a solution to this problem, Crocosphaera performs photosynthesis during the day and nitrogen fixation at night.

During the BioLINCS cruise, researchers will be studying *Crocosphaera* using the influx flow cytometer (IFC).

Uncultivated Cyanobacteria Group A (UCYN-A)

UCYN-A is a group of nitrogen-fixing cyanobacteria that were first discovered in the open ocean near Hawaii in 1998. Most of the time, they are not the most abundant nitrogen-fixing organism, but they can reproduce rapidly when conditions are right.

UCYN-A bacteria contain very few genes, compared with most other marine cyanobacteria. Although the entire UCYN-A genome has been sequenced, the organisms themselves have yet to be grown in the laboratory.

Surprisingly, sequencing of the UCYN-A genome revealed that that the bacteria lack genes that are necessary for "fixing" inorganic carbon and for key parts of the photosynthetic process. This allows UCYN-A to fix nitrogen during the daytime. However, because the bacteria cannot perform photosynthesis, they must have some other means of acquiring key organic compounds that are normally produced during photosynthesis. As with some heterocystus cyanobacteria, it is possible that UCYN-A form "partnerships" (symbioses) with other photosynthetic organisms to acquire these essential compounds.

During the BioLINCS cruise, UCYN-A are being studied using DNA and RNA analyses that are carried out manually on board the ship and performed automatically within the Environmental Sample Processor (ESP).

Alphaproteobacteria and Gammaproteobacteria

Proteobacteria are an extremely diverse group of bacteria. Some alphaproteobacteria and gammaproteobacteria, two subgroups within the proteobaceria, carry the nifH gene, and may be able to fix nitrogen. The proteobacteria also include some key players in the process of "nitrification," as described below.

During the BioLINCS cruise, the Environmental Sample Processor (ESP) will be looking for the nitrogenase (nifH) gene specific to gammaproteobacteria. This gene allows the bacteria to produce the enzyme nitrogenase, which they use to "fix" nitrogen. The ESP will also be looking for alphaproteobacteria using sandwich hybridization arrays.

Marine Bacteria Involved in Nitrification

Ammonium Oxidizing Bacteria (AOB)

The ammonium oxidizing bacteria (AOB) are bacteria that are all involved in a specific biochemical process (nitrification), but which may or may not be related to one another. They are involved in the first step of nitrification—the conversion of ammonium to nitrite (also known as "ammonium oxidation").

Most of the known ammonium oxidizing bacteria is betaproteobacteria and gammaproteobacteria. These bacteria are believed to be involved in this process because they carry a gene that codes for the ammonia monooxygenase enzyme (*amoA*).

Nitrite Oxidizing Bacteria

Several different genera of marine bacteria are involved in in the second step in the nitrification process—converting nitrite to nitrate (also known as "nitrite oxidation"). These include *Nitrobacter*, *Nitrospira*, and *Nitrospina*. Of these, *Nitrospira* is believed to be the most important in the open ocean.

Marine bacteria that are primary producers but do not fix nitrogen.

Prochlorococcus

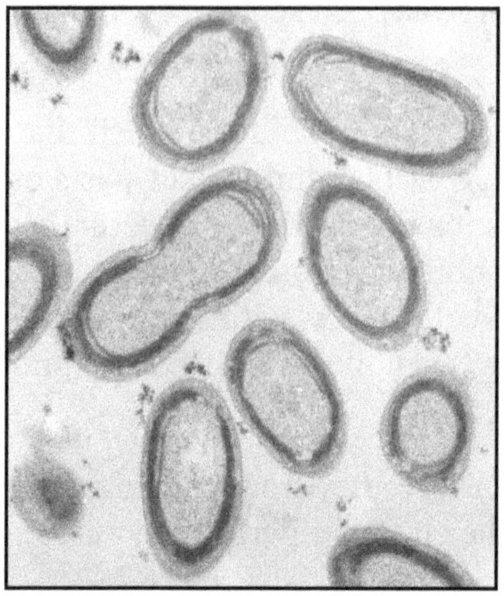

This scanning electron microscope

Prochlorococcus is a genus of cyanobacteria that is very common in open-ocean areas around the world. Although extremely tiny, with cells only 0.5 to 0.8 microns across, *Prochlorococcus* (along with another genus, *Synechococcus*) are so widespread and abundant that may produce a third of the oxygen in the Earth's atmosphere. *Prochlorococcus* are "obligate photoautotrophs," obtaining all of their energy through photosynthesis. However, they require nitrogen as a nutrient, and can use either nitrate or ammonium as a source of nitrogen. Thus, they are a key part of the "assimilation" process in oceanic nitrogen cycles.

During the BioLINCS cruise, researchers will be studying *Prochlorococcus* using the influx flow cytometer (IFC), and in some on-deck incubation experiments.

Synechococcus

Synechococcus is another common type of marine cyanobacteria. It is arguably the second most common group of photosynthetic marine bacteria, after Prochlorococcus. It has a similar shape as Prochlorococcus, but is typically a little bit larger, at 0.8 to 1.5 microns across.

Like Prochlorococcus, Synechococcus are "obligate phototrophs," which means that they can only obtain energy through photosynthesis, and require nitrate as a key nutrient. They can use nitrate or ammonium as sources of nitrogen, and are a key part of the "assimilation" process in oceanic nitrogen cycles.

During the BioLINCS cruise, researchers will be studying *Synechococcus* using the influx flow cytometer (IFC).

Marine bacteria play a crucial role in biogeochemical processes such as the cycling of carbon and nitrogen. They have a variety of functions, which in some species are triggered only when seasonal conditions indicate that the time is right.

Marine bacteria provide a useful source for bioremediation and have an advantage over terrestrial bacteria because they can be used in extreme environmental conditions. As the bacteria form biofilms, there is a greater increase in their bioremediation capability due to the presence of EPS like polysaccharides, amyloids, extracellular enzymes, and bio surfactants. In this regard, the isolated marine biofilm-forming mercury-resistant bacterium B. cereus BW-03 holds promise as a suitable candidate species and as an alternative for current remediation of inorganic mercury and mixed waste. It can be suggested that the biosorption aspect of mercury bioremediation should be given priority over the volatilization aspect in future.

Marine Archaea

Archaea are ubiquitous and abundant members of the marine plankton. Archaea are single celled organisms that look similar to bacteria, but which are in an entirely separate biological domain. Historically, archaea were thought of as living mostly in extremely hot, acidic, or low oxygen environments. However, in the last decade, using DNA and RNA analysis, molecular biologists have found them to be very common in both freshwater and saltwater environments.

Ammonium Oxidizing Archaea (AOA)

Like ammonium oxidizing bacteria, ammonium oxidizing archaea are important in the nitrification process in open-ocean areas. They perform the first step of nitrification—the conversion of ammonium to nitrite, which is also known as "ammonium oxidation." In the process of performing ammonium oxidation process, AOA may produce nitrous oxide, a greenhouse gas that is sometimes released from open-ocean waters into the atmosphere.

All of the known ammonium oxidizing archaea are in the group *Thaumarchaea* (formerly known as *Crenarchaeota*). These archaea are common in soils, in estuaries, and in the deeper parts of the open ocean, where there is little light and oxygen concentrations are relatively low.

Like ammonium oxidizing bacteria (AOB), AOA are able to make an enzyme called ammonia monooxygenase. Researchers often use DNA analysis to determine if a sample of seawater contains the gene (*amoA*) that allows AOA and AOB to produce ammonia monooxygenase. Once they identify organisms with this gene, the researchers can use genetic probes to tell if the organisms are bacteria or archaea.

Eukaryotes

Marine Protists

The ocean is teeming with a fascinating world of organisms the marine protists. With such a huge variety of organisms, there are some really interesting marine protists. Protists are organisms in the kingdom Protista. These organisms are eukaryotes, meaning they are made up of single or multiple cells which all contain a nucleus enclosed by a membrane. The protists are a diverse group of eukaryotes that cannot be classified as animals, plants, or fungi. Organisms in the Protista kingdom include amoebae, red algae, din flagellates, diatoms, euglena, and slime molds.

Protists are defined by how they obtain nutrition and how they move. Protists are typically divided into three categories, including animal-like protists, plant-like protists, and fungus-like protists.

Protists vary in how they move, which can range from cilia, flagella, and pseudopodia. In other words, protists move by microscopic hair that flaps together, by a long tail that moves back and forth, or by extending its cell body, similar to an amoeba.

Nutritionally, protists tend to gather energy in a variety of ways. They can either eat food or digest it inside of themselves, or they may digest outside of their bodies by secreting enzymes. Other protists, like algae, perform photosynthesis and absorb energy from sunlight to make glucose.

Animal-Like Protists

Some protists look like animals and are typically referred to as protozoa. Most of these types of protists are made up of a single cell and are similar to animals in nature because they are heterotrophs and able to move around. While they are not considered animals themselves, it is often thought that they may be a shared ancestor. Examples of animal-like protists include:

- Zooflagellates – flagella
- Sarcodines – extensions of cytoplasm (pseudopodia)
- Ciliates – cilia
- Sporozoans

Plant-Like Protists

There is also a large and diverse group of protists that are plant-like and known as algae. While some are single-celled, others like seaweed have multiple cells. For example, one type of protist in the marine environment is Irish moss, which is a species of red algae.

More plant-like protists include:

- Dinoflagellates
- Diatoms
- Euglenoids
- Red algae
- Green algae
- Brown algae

Fungus-Like Protists, Lastly, there are fungus-like protists that are also known as molds. These feed on dying organic matter and look like fungi. The major protists in this family include slime molds and water molds. Slime molds can be found on rotting logs and compost while water molds are seen in moist soils and surface waters. Examples of fungus-like protists may include:

- Dictyosteliomycota
- Myxomycota
- Labyrinthulomycota
- Oomycetes

Benefits

Protists are important to the world in several ways. You may be surprised to learn that chalk is made from the fossil shells of protists, which is helpful in our classrooms and our children's creativity and play. Additionally, protists produce oxygen which is helpful for the planet.

Many protists have a high nutritional value which can help improve illnesses. Protists like protozoa are used in foods like sushi and are good for our water, as protozoa are used to prey on bacteria and help to clean water for us to use.

Marine Microalgae

The primary producers of oxygen in aquatic environments are algae, especially planktonic microalgae. These microorganisms are widely distributed in nature and have adapted to different environments with great diversity in size, morphology, life cycle, pigments, and metabolism. About one half of global photosynthesis and oxygen production is accomplished by marine microalgae. They play an important role in CO_2 recycling through photosynthesis, which is similar to higher plants in O_2-evolved systems (PSI and PSII). Research in microalgae has been carried out not only on physiological aspects but also to develop production of useful biomaterials. The advantages of their utilization in production are (1) their ability to convert CO_2 to useful materials through photosynthesis and (2) their ability to grow in natural environments under inorganic conditions. For example, marine microalgae can be cultivated using seawater, CO_2, and sunlight. Recent developments in the biotechnology of microalgae have been focused on their production of useful materials applicable to the cosmetic and medical fields.

Genetic modification and molecular tools have been developed mainly in eubacterial microalgae, cyanobacteria (blue-green algae). In contrast, genetic modification has been only gradually applied to eukaryotic microalgae. Recently, whole genome sequences and EST analyses have been performed in marine strains. The elucidation of genomewide information may help in the development of new biotechnological applications using microalgae.

In this chapter, we review the useful applications of microalgae for genetic engineering, cultivation technologies, and CO_2 fixation as follows:

- Production of useful chemicals by marine microalgae
- Metabolic engineering of marine microalgae
- Microalgal mass cultivation technologies
- CO_2 fixation using microalgal cultures in industry

Production of useful Chemicals by Marine Microalgae

Cyanobacteria

The cyanobacteria are oxygenic photosynthetic prokaryotes that show large diversity in their morphology, physiology, ecology, biochemistry, and other characteristics. Currently, more than 2000 species are recognized, which comprise two fifths of known bacterial species (5000 species). Such variation complicates estimation of species diversity. Cyanobacteria contain chlorophyll *a*, phycobiliproteins, and characteristic glycosylated xanthophylls, such as myxoxanthophyll and oscillaxanthin. The phycobiliproteins are water- soluble pigments consisting of red-colored phycoerythrin and blue colored phycocyanin and allophycocyanin. They have a characteristic structural feature, the phycobilisome that is used as a photosynthetic light-harvesting antenna. Cyanobacteria are unicellular, multicellular, colonial, and branched or unbranched filamentous forms. Specialized cells, heterocystous and akinates, are contained in some of the filamentous-form cells. Some cells produce ex- tracellular matrix such as sheaths, capsules, and slimes that consist mainly of polysaccharides. Cyanobacteria are distributed widely not only in salt water but also freshwater, brackish water, polar region, hot springs, and deserts. Some also exist as symbionts in sponges, ascidians, echiuroid worms, plank- tonic diatoms, and dinoflagellates in marine environments and lichens and azollae in terrestrial environments. Cyanobacteria, especially marine pelagic *Synechococcus* and *Prochlorococcus*, contribute largely to global oxygen production.

Many commercial applications have been proposed for marine cyanobacteria, although no marine strain presently is commercially supplied. Re- cent proposals to evaluate potential commercial uses typically fall into three categories: bioactive chemical compounds, polysaccharides, and evaluation of new genes for recombinant expression. Cyanobacteria can produce a large variety of complex chemical compounds. Bioactive com- pounds isolated from marine cyanobacteria have recently been summarized by Burja et al. and Takeyama and Matsunaga. Novel plant growth regulators that promote redifferentiation, germination, and plantlet formation, tyrosinase inhibitors, UV-A absorbing compounds, sulfated polysaccharides showing anti-HIV activity, and novel antibiotics with light-regulated activity are among the many compounds that have been studied. Matsunaga et al. reported that several marine cyanobacterial strains, such as *Phormidium* sp. NKBG 041105 and *Oscillatoria* sp. NKBG 091600, showed high *cis*-palmitoleic acid content (54.5% and 54.4% of total fatty acid, respectively).

The discovery of biochemically active compounds from marine cyanobacteria, including enzyme inhibitors, herbicides, antimycotics, antifeedants, multi-drug-resistance reversers, and antimalarial and immune supressive agents, has dramatically increased over the last few years. This has been due to the adaptation and use of current cyanobacteria collections and cyanobacteria-derived compounds for screening in new pharmaceutical and industrial assays.

The focus on the polysaccharides of marine cyanobacteria as well as fresh- water strains also has greatly increased in relation to interest in their exploitation for various industrial applications. Cyanobacteria produce three types of extracellular matrix consisting mainly of polysaccharides, which have unique bio- and physicochemical characteristics. Most of them are composed of at least ten different monosaccharides and contain pentoses, which have not been observed in other prokaryotic polysaccharides. The anionic nature of these unique polysaccharides is due to the presence of acidic sugars and anionic organic and inorganic compounds. Little work has been devoted to potential applications of marine cyanobacterial polysaccharides.

Marine cyanobacteria also are attractive as a resource for useful enzymes and genes. Cyanobacteria commonly produce complex macro- molecules that often possess biological activities such as cytotoxicity or microbial toxicity. Recent genetic analyses have revealed that most of these macromolecules, as well as unusual small molecules, are coded for by large gene clusters representing nonribosomal peptide synthetase and polyketide synthetase. These gene clusters may be manipulated for the production of new chemicals. Many varieties of the gene clusters are present in most cyanobacteria. Therefore, cyanobacteria are attractive not only as producers of useful bioactive macromolecules and enzymes but also for production of complex macromolecules that may become import- ant pharmaceuticals.

Rhodophyta

The rhodophytes, or red algae, contain chlorophyll *a*, carotenoids, and phycobiliproteins. Their red color is due to the presence of phycoerythrin in the outermost part of the phycobilisomes, while other regions of the algae look blue-green due to the absence of phycoerythrin. Rhodophytes are unicellular, or composed of simple or complex filamentous aggregates. Flagellated cells have not been observed. Red macroalgae commonly inhabit tropical and temperate zones near shores. About 600 genera and 5500 species have been recognized. Most of them (98%) are marine macroalgal species.

Red macroalgae are of economic significance. *Porphyra* and a few other species are cultured for human food. The production of red algal polysaccharides such as agar, agarose, and carrageenans is also an important industry. These compounds are widely used for laboratory cell culture media, nucleic acid purification, or food processing, respectively.

Although rhodophyte microalgae have not been produced commercially as yet, their polysaccharides are considered to have commercial potential. For example, in concentrated solutions of polysaccharides from the unicellular rhodophyte *Porphyridium* sp., viscosity is stable over a wide range of pH, temperature, and salinity. These properties indicate possible applications for use as a thickening agent in aqueous systems or as a stabilizer for emulsions and suspensions. In addition to its potential application as a viscosity stabilizer, the polysaccharide and the biomass of *Porphyridium* sp. have been used as functional food additives. The colon and jejunum changed morphologically with hypertrophy in the muscularis layer in rats fed diets containing pelleted biomass or sulfated polysaccharides. In addition, it was shown in rats that the algal polysaccharide and biomass were potent hypocholesterolemic agents active at low concentrations in the diet. More- over, the sulfated polysaccharide of *Porphyridium* sp. has shown promising antiviral activity against a variety of animal viruses including *Herpes simplex* viruses types 1 and 2 (HSV 1, 2) *Varicella zoster* virus (VZV), and HIV types 1 and 2. The compounds also showed significant inhibition of productive infection with retroviruses (murine leukenia virus, HIV-1, and HIV-2) and cell transformation by murine sarcoma virus in vitro. Thus, red microalgae and their polysaccharides seem to be good candidates for the development of antiviral drugs.

Chlorophyta

Cells of chlorophytes are green due to chlorophyll *a* and *b*, the same pre- dominant photosynthetic pigments as those of land plants. Some algae show yellowish-green or red-green colors due to the presence of a certain amount of carotenoids such as β-carotene, prasinoxanthin, siphonaxanthin, and as-taxanthin. These chlorophyta form starch in the chloroplast as a storage product of photosynthesis. Chlorophytes are unicellular, multicellular, colonial, filamentous, siphonous, and thallus. The Chlorophyta consist of five classes, the Prasinophyceae, the Ulvophyceae, the Chlorophyceae, the Trebouxiophyceae, and the Charophyceae. The Treboxiophyceae recently were separated from the Chlorophyceae. The Chlorophyta are primarily freshwater algae with approximately 500 genera comprising 16 000 species. Only about 10% of these are marine species. The Ulvophyceae are primarily multicellular marine green algae. In addition, some species from the Prasinophyceae, Chlorophyceae, and Trebouxiophyceae families are found in the marine environment. Depending on the class, it has been estimated that there would be at least two to three times more species in this phylum.

A marine species of the Chlorophyceae, *Dunaliella*, has been cultivated commercially for food supplements and β-carotene production. The microalgal biomass of some marine *Tetraselmis* and *Pyramimonas* strains in the Prasinophyceae family also are used for fish food additives. Recently, anti-inflammatory and immunosupressive properties were discovered in the extracts or extracted polysaccharides of another marine species, *Chlorella stigmatophora*. Miura et al. reported that *Chlorella* sp. NKG 042401 contains 10% γ -linolenic acid (C18 : 3), which is present in the cells mainly in the form of galactolipid.

Cryptophyta

Cryptophytes are unicellular, cryptomonad flagellates with 12 to 23 genera comprised of 200 species. A few species are colorless heterotrophs, but most possess various colored plastids with chlorophylls *a* and *c*, carotenoids, and phycobiliprotein. Alloxanthin is a xanthophyll that is unique to cryptomon- ads. Morphologically, cells have a flattened asymmetrical shape with two anterior flagella, slightly unequal in length. They are distributed widely both in freshwater and marine environments. Approximately half of the known species inhabit marine environments. It has been estimated that there would be about six times more species in this phylum.

Only a few strains such as *Rhodomonas minuta* and *Cryptomonas* sp. have been used for aquaculture feeds since they contain significant amounts of polyunsaturated fatty acids (PUFAs). No further application has been proposed.

Heterokontophyta

The phylum Heterokontophyta is the most diverse algal group with huge commercial and biotechnological potentials. They range in size from microscopic unicells to giant kelp averaging several meters. Cells of heterokontophytes contain chlorophyll *a* with chlorophyll c and carotenoids such as fucoxanthin or vaucheriaxanthin. They are characterized primarily by the similarities in their ultrastructural and biochemical characteristics.

Heterokontophyte microalgae are widely used as feed in mariculture/aquaculture. Diatoms such as *Chaetoceros calcitrans*, *Chaetoceros gracilis*, *Chaetoceros muelleri*, *Skeletonema costatum*, and *Thalasiosira pseu- dodonana* are commonly used as live feeds for all growth stages of bivalve molluscs (e.g., oysters, scallops, clams, and mussels), for crustacean larvae, and for zooplankton used as feed for larvae. The genera *Navicula*, *Nitzschia*, *Cocconeis*, and *Amphora* also are used to feed juvenile abalone. Some Eustigmatophyceae species of the genus *Nannochloropsis* are commonly fed to Artemia or rotifers, which in turn are fed to crustacean and fish larvae.

The biotechnological potential of diatoms is also concerned with PUFA production. Most diatoms have a high content of eicosapentenoic acid (EPA) 20:5 (n-3). *Phaeodactylum tricornutum* and *Nitzschia laevis* especially have been investigated for EPA production. In addition, EPA production by diatoms has been reviewed recently by Lebeau and Robert. Recent advances in heterotrophic production of EPA by microalgae were also reviewed by Wen and Chen.

The Pinguiophyceae also have significant biotechnological potential for use in fish feed and for PUFA production . Pinguiophyceae consist of five marine unicellular algal species, *Pinguiochrysis pyriformis*, *Phaeomonas parva*, *Pinguiococcus pyrenoidosus*, *Glossomastix chrysoplasta*, and

Polypodochrysis teissieri. These algae have an unusually high percentage of PUFAs, especially EPA. The EPA content ranges from 23.5% to 56.0% of the total fatty acids in these five species. They also contain arachidonic acid (AA) 20:4 (n-6) and docoahexaenoic acid (DHA) 22:6 (n-3). *Pinguiococcus pyrenoidosus, Glosso- mastix chrysoplasta,* and *Polypodochrysis teissieri* have docosatetraenoic acid (DTA) 22:4 (n-3) ranging from 4.4% to 9.5% of total fatty acid content. This significant oil balance, as well as their lack of cell wall, indicates that they are good candidates as food/feed sources. Evaluations of the utility of Pinguio- phyceae for food/feed supplements and optimization of growth conditions necessary for efficient production are being carried out.

Some marine colorless heterokontophytes also are being tested for the production of DHA. *Schizochytrium, Thraustochytrium,* and *Ulkenia* are representatives of the class Labyrinthulea. They produce substantial amounts of PUFAs, especially DHA. The safety of biomass production and differences in oil content are intensively studied. Utilization of fermented "okara" for DHA and EPA production by thraustochitrids also has been reported, while the yield was lower by growth in a glucose yeast-extract medium than by fermentation.

Dinophyta

The Dinophyta include the dinoflagellates, most of which are unicellular, with two dissimilar flagella. Flagellated cells show characteristic forward-spiraling swimming motions. Organisms in this phylum have remarkable morphological diversity including nonflagellate amoeboid, coccoid, palmelloid, or filamentous. Approximately 130 genera with about 2000 living and 2000 fossil species have been described in this group. Most are marine and only about 220 species are from freshwater. About half of the known species are colorless heterotrophs. Most of the plastid-containing phototrophic dinoflagellates contain chlorophylls *a* and c_2 and carotenoids such as β-carotene and peridinin, the unique accessory xanthophyll in this phylum. Dinoflagellates are also characterized by cell coverings consisting of a layer containing many closely adjacent, flattened amphiesmal (thecal) vesicles. In many species, each amphiesmal vesicle contains a thecal plate composed of cellulose. About 60 dinoflagellate species are known to pro- duce cytolytic, hepatotoxic, or neurotoxic compounds. Some form harmful red tides. The majority of toxin-producing dinoflagellates are photosynthetic, estuarine, or coastal shallow-water forms that are capable of producing benthic resting cysts and that tend to form monospecific populations. Fresh- water species are not known to produce toxins. The dinophytes comprise autotrophs, mixotrophs, osmotrophs, phagotrophs, and parasites. Dinoflagellate endosymbionts known as zooxanthellae are essential for the existence of coral reef ecosystems. Symbiotic as well as parasitic forms also are present within the cells or tissues of fish, invertebrates, and filamentous algae.

Biotechnological applications for dinoflagellates have not been intensively performed, probably because most dinoflagellates can not be easily cultured. The only exception is the DHA produced by heterotrophically grown *Crypthecodinium cohnii.* DHA produced by these cells is distributed widely as food supplements, such as for infant formula. Recently, significant biological activities have been attributed to the polysaccharide from *Gymnodinium* sp. For example, strong cytotoxicity against several human leukemic cell lines leading to apoptosis and potent anticancer activity mediated by the inhibition of topoisomerase I and II. Optimal growth conditions for *Gyrodinium impudicum* were reported to produce a sulfated polysaccharide that showed antiviral activity against encephalomyocarditis virus.

Haptophyta

The phylum Haptophyta (haptophytes or prymnesiophytes) is a group of unicellular flagellates characterized by the presence of a haptonema between two smooth flagella. The role of haptonema is unclear, although it some- times functions as a feed-capturing net, in avoidance reaction by coiling and recoiling, and as an attachment organ on surfaces. There also are haptonema- less flagellates or nonflagellate unicells or colonies. The cells of haptophytes are brownish or yellowish-green containing chlorophylls *a* and c_1/c_2 and carotenoids such as β-carotene, fucoxanthin, diadinoxanthin, and diatoxanthin. The cells are commonly covered with scales made mainly by carbohydrates or calcium bicarbonate. Many species known as coccolithophorids produce calcified scales called coccoliths. About 70 genera with 300 species have been recognized to date. Most are primarily marine species inhabiting tropical seawater. The group is distributed worldwide and is often an important source of food for aquatic communities. Some haptophytes, however, produce algal blooms and cause serious problems for fish and for fishermen by producing dimethyl sulfide (DMS), a noxious-smelling compound that affects fish migrations and alters their normal routes.

Microalgal biomass of haptophytes is commonly used as living feed in aquaculture. *Isochrysis galbana* and *Pavlova lutheri*, especially, are used as living feeds for bivalve molluscs, crustacean larvae and zooplanktons that in turn are used for crustacean and fish larvae. Some cells can produce PUFAs such as DHA or EPA. In addition, the DHA content in *I. galbana* was shown to be enhanced by low temperature or incubation of the culture in the dark after reaching plateau phase growth. Furthermore, it was shown that these algae are useful for DHA enrichment of feed such as rotifers for the larvae of several marine fish species.

Euglenophyta

Euglenophytes are unicellular organisms with two pantonematic flagella arising from the bottom of a flask-shaped invagination called a "gullet." A few have stages with colonies or are enclosed within a mucilaginous capsule. There are about 40 genera comprised of 900 species, of which one third have green plastids. The chloroplast originating from green algae contains chlorophylls *a* and *b* and carotenoids such as diadinoxanthin, neoxanthin, and β-carotene. Two thirds of the genera are heterotrophic, some having colorless plastids and some lacking plastids altogether, living either saprotrophically or phagocytically. Cell walls are absent, but there is a characteristic cell-covering structure called a pellicle, composed of protein-rich spiral strips beneath the cell membrane. One to several flagella may be present, and nonflagellate cells can undergo a type of motion involving changes in cell shape. Although most species are found in highly eutrophic freshwater environments such as ponds and ditches, some euglenoids have important ecological roles in particular marine environments. Very few euglenoids have been grown in axenic culture, and euglenoid culture media are generally very nutrient rich. No direct economic significance has been associated with this phylum, probably because of the difficulties in culturing. Although euglenoids generally are harmless, toxin production has been demonstrated in some freshwater *Euglena* sp. On the other hand, *E. gracilis Z* is one of the few microorganisms that simultaneously produces antioxidant vitamins such as β-carotene and vitamins *C* and *E*. Efficient production of these vitamins consists of a two-step culture under photoheterotrophic/photoautotrophic conditions.

Phytoplankton

Phytoplankton, provide the food base which supports the entire animal population of the open sea. Most algal classes are represented in ocean populations: the best known are the diatoms (Bacillariophyta), the dinoflagellates (Dinophyta), the green algae (Chlorophyta) and the blue-green algae (Cyanophyta). They are the autotrophic component of the plankton community. As plankton, they are organisms (mostly microscopic) that drift in the oceans, lakes, rivers, and other bodies of water. As autotrophs, they are the producers in a food chain, producing complex, energy-rich organic compounds, such as carbohydrates, from simple inorganic molecules using energy from light (photoautotrophs), or, less commonly, energy from inorganic chemical reactions (chemoautotrophs).

Most phytoplankton are too small to be individually seen with the unaided eye. However, when present in high enough numbers, they may appear as a green discoloration of the water due to the presence of chlorophyll within their cells (although the actual color may vary with the species of phytoplankton present due to varying levels of chlorophyll or the presence of accessory pigments such as phycobiliproteins, xanthophylls, etc.)

Phytoplankton provides an essential ecological function for all aquatic life by serving as the base of the aquatic food web. They also provide a function for almost all life on earth, as it is phytoplankton that is responsible for much of the oxygen present in the Earth's atmosphere. Phytoplankton also serves as a key food item in both aquaculture and mariculture.

Phytoplankton bloom in the South Atlantic seen from space

The name phytoplankton comes from the Greek words *phyton*, meaning "plant," and *planktos* meaning "wanderer" or "drifter". These organisms are carried along by the water current, in contrast to nekton organisms that can swim against the current and control their position, and in contrast to benthic organisms that live on the bottom of bodies of water.

Phytoplankton is autotrophs in that they are able to take energy from the environment in the form of sunlight or inorganic chemicals and use it to create energy-rich molecules such as carbohydrates. This mechanism is called primary production. It is in contrast to heterotrophs, which take in autotrophs as food to carry out functions necessary for their life. Thus, heterotrophs depend on autotrophs for the energy and raw materials they need.

Phytoplankton most commonly obtains energy through the biological process called photosynthesis. They therefore must live in the well-lit surface layer (termed the euphotic zone) of an ocean, sea, lake, or other body of water. Through photosynthesis, phytoplankton are responsible for much of the oxygen present in the Earth's atmosphere—half of the total amount produced by all plant life.

In addition to these photoautotrophs, which carry out photosynthesis to acquire energy (using sunlight, carbon dioxide, and water), there are also chemoautotrophs. Chemoautotrophs do not use solar energy as their energy source, but rather energy by the oxidation of electron-donating molecules in their environment. They do synthesize all necessary organic compounds from carbon dioxide, rather than organic sources of carbon. Examples include bacteria that derive energy from oxidizing inorganic compounds such as hydrogen sulfide, ammonium and ferrous iron. Organisms that require organic compounds as a source of carbon, even if they use light or inorganic compounds as a source of energy, are not defined as autotrophic, but rather as heterotrophic.

In addition to sunlight (or inorganic energy sources in the case of chemoautotrophs) and inorganic carbon sources, phytoplankton are also crucially dependent on minerals. These are primarily macronutrients such as nitrate, phosphate, or silicic acid, whose availability is governed by the balance between the so-called biological pump and upwelling of deep, nutrient-rich waters. However, across large regions of the World Ocean such as the Southern Ocean, phytoplankton is also limited by the lack of the micronutrient iron. This has led to some scientists advocating iron fertilization as a means to counteract the accumulation of human-produced carbon dioxide (CO_2) in the atmosphere. Large-scale experiments have added iron (usually as salts such as iron sulphate) to the oceans to promote phytoplankton growth and draw atmospheric CO_2 into the ocean. However, controversy about manipulating the ecosystem and the efficacy of iron fertilization has slowed such experiments.

Phytoplankton's cumulative energy fixation in carbon compounds (primary production) is the basis for the vast majority of oceanic and also many freshwater food webs. As a side note, one of the more remarkable food chains in the ocean—remarkable because of the small number of links—is that of phytoplankton fed on by krill (a type of shrimp) fed on by baleen whales.

Types of Phytoplankton

The term phytoplankton includes all planktonic, photoautotrophic microorganisms in aquatic food webs. However, unlike terrestrial communities, where most autotrophs are plants, phytoplankton are a diverse group, incorporating protistan eukaryotes and both eubacterial and archaebacterial prokaryotes. While almost all phytoplankton species are obligate photoautotrophs, both photoautotrophs and chemoautotrophs are included in the phytoplankton. Overall, there are about 5,000 described species of marine phytoplankton. There is uncertainty in how such diversity has arisen in an environment where competition for only a few resources would suggest limited potential for niche differentiation.

In addition, the plankton in general includes zooplankton (heterotrophic, sometimes detritivorous, type of plankton). There also are organisms that are called mixotrophic because they obtain their electrons from an inorganic electron source, such as hydrogen-sulfide, ammonium, and hydrogen, while using organic matter as a carbon source. These later are sometimes labeled as phytoplankton although they technically are heterotrophs, since their carbon source is not inorganic but organic. An organism that obtains carbon from organic compounds but obtains energy from light is called a photoheterotroph, while an organism that obtains carbon from organic compounds but obtains energy from the oxidation of inorganic compounds is termed a chemoheterotroph. Among non-pigmented species that are sometimes known as phytoplankton, but actually are heterotrophic and generally viewed as zooplankton, are dinoflagellate genera such as *Noctiluca* and *Dinophysis*; that obtain organic carbon by ingesting other organisms or detrital material.

In terms of numbers, the most important groups of phytoplankton include the diatoms, cyanobacteria, and dinoflagellates, although many other groups of algae are represented. One group, the coccolithophorids, is responsible (in part) for the release of significant amounts of dimethyl sulfide (DMS) into the atmosphere. DMS is converted to sulfate and these sulfate molecules act as cloud condensation nuclei, increasing general cloud cover. In oligotrophic oceanic regions such as the Sargasso Sea or the South Pacific gyre, phytoplankton is dominated by the small sized cells, called picoplankton, mostly composed of cyanobacteria (*Prochlorococcus*, *Synechococcus*) and picoeucaryotes such as *Micromonas*.

Diatoms

Diatoms

Diatoms are one of the most common types of phytoplankton and are found in large numbers in almost all waters of the world, freshwater and saline (as well as in moist soil and other environments). Diatoms are unicellular or (less commonly) colonial algae characterized by unique cell walls of silica. In favorable conditions, these protists "bloom," reproducing rapidly so that they dominate an area. When conditions turn less favorable, for instance the coming of winter or a depletion of nutrients, they can turn dormant and sink to lower levels until currents return them to the surface and a new bloom can take place.

Dinoflagellates

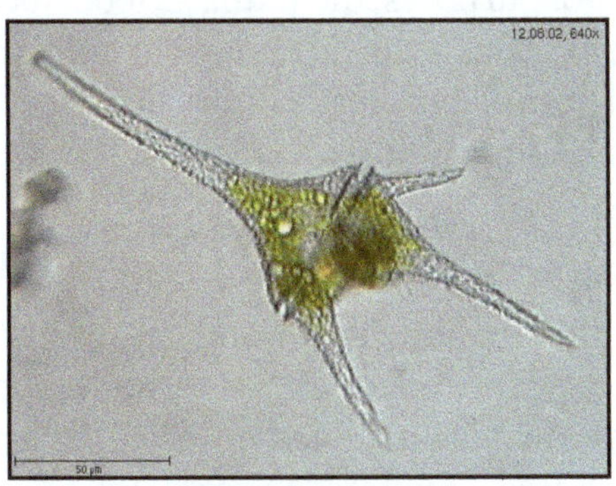

Dinoflagellate

Dinoflagellates are also protists, but differ from diatoms in having a whip-like tail (flagellum) that enables them to move in the water, mostly up and down in response to changing conditions. They typically are characterized by being single-celled and with two dissimilar flagella during at least part of their life cycle. Among the dinoflagellates are both species exhibiting the characteristics of plants and those exhibiting characteristics of animals, with about half of them being photosynthetic autotrophs, while the balance are heterotrophs securing nutrients by acting as predators, endosymbionts, or parasites. Most dinoflagellates are marine plankton, but they are common in fresh water habitats as well. Dinoflagellates also bloom, which can create a "red tide" in which they release toxins that can kill fish and humans who eat shellfish in which the toxins have become concentrated. Those dinoflagellates that do not use photosynthesis but eat other organisms are considered zooplankton.

Cyanobacteria

Cyanobacteria, also called "blue-green algae," are bacteria. They are thought to be the first organisms on earth that developed photosynthesis. They are one of the few organisms that can take nitrogen from the air and "fix" it into organic molecules so it can be used by plants and animals.

Seaweeds

Although the word "plankton" is most often used for microscopic, or at least very small organisms, seaweeds, which are multi-cellular algae, may be considered planktonic if they drift in the open ocean rather than are anchored to the bottom. Sargassum weed forms huge floating islands in the western Atlantic Ocean that provide a home for many unique animals.

Aquaculture

Phytoplankton are a key food item in both aquaculture and mariculture. Both utilize phytoplankton for the feeding of the animals being farmed. In mariculture, the phytoplankton is naturally occurring and is introduced into enclosures with the normal circulation of seawater. In aquaculture, phytoplankton must be obtained and introduced directly. The plankton can either be collected from a body of water or cultured, though the former method is seldom used.

Phytoplankton is used as a foodstock for the production of rotifers, which are in turn used to feed other organisms. Phytoplankton is also used in many varieties of aquacultured mollusks, including pearl oysters and giant clams.

The production of phytoplankton under artificial conditions is itself a form of aquaculture. Phytoplankton is cultured for a variety of purposes, including as foodstock for other aquacultured organisms, a nutritional supplement for captive invertebrates in aquaria, and as a source of bio-diesel. Culture sizes range from small-scale laboratory cultures of less than one liter to several tens of thousands of liters for commercial aquaculture.

Regardless of the size of the culture, certain conditions must be provided for efficient growth of plankton. The majority of cultured plankton is marine, and seawater of a specific gravity of 1.010 to 1.026 may be used as a culture medium. This water must be sterilized, usually by either high temperatures in an autoclave or by exposure to ultraviolet radiation, to prevent biological contamination of the culture. Various fertilizers are added to the culture medium to facilitate the growth of plankton. A culture must be aerated or agitated in some way to keep plankton suspended, as well as to provide dissolved carbon dioxide for photosynthesis. In addition to constant aeration, most cultures are manually mixed or stirred on a regular basis. Light must be provided for the growth of phytoplankton. The color temperature of illumination should be approximately 6,500 K, but values from 4,000 K to upwards of 20,000 K have been used successfully. The duration of light exposure should be approximately 16 hours daily; this is the most efficient artificial day length.

Marine Fungi

Marine fungi are an ecologically diverse group which belong to the phyla Aphelidiomycota, Ascomycota, Basidiomycota, Blastocladiomycota, Chytridiomycota, and Mucoromycota. They grow on numerous substrata such as decaying wood and leaves, algae, coral, calcareous tubes of molluscs, animals, and found in sand, muds, soils, sediments. Marine fungi play a substantial components role in nutrient cycling and are a critical source of natural products. The Halosphaeriaceae is the largest family of marine species with 141 species in 59 genera. The Aphelidiomycota currently includes one marine species Pseudoaphelidium drebesii.

Sources of Marine Fungi

Filamentous fungi are major decomposers of woody and herbaceous substrata entering marine ecosystems and propagate by spores, either asexual conidia or by sexual ascospores, basidiospores, or zygospores. In marine ecosystems, spores are dispersed and passively transported through water, ultimately landing on a suitable substrate where they germinate and colonize. During colonization, hyphae (fungal filaments) grow throughout the substrate and, when environmental conditions are appropriate, differentiate into specialized structures to reproduce. Observations of fruiting structures, such as pycnidia or ascomata (typically observed as round "balls" or "tiny dots" emerging from or forming on the substrate surface) and conidiophores ("branch-like" structures bearing wet or dry masses of conidia), are sought after by mycologists for isolating and identifying marine fungi.

Figure: Fruiting structures of marine fungi

(a) Amylocarpus encephaloides is one of the few species of marine fungi that can be spotted and identified with the naked eye already in the field; (b) these "black dots" on the receptacle of macroalga Ascophyllum nodosum are fruiting bodies of Stigmidium ascophylli (syn. Mycosphaerella/ Mycophycias ascophylli); (c) apothecial fruiting bodies of Calycina marina on dead and decaying A. nodosum; (d) conidial mass of the asexual form of the fungus, Lulwoana uniseptata (=Zalerion maritima), in a borehole made by a Xylophaga bivalve species.

Marine fungi are found on a diverse range of substrates: Submerged wood, estuarine plants (such as mangroves and salt marsh grasses and rushes), sand grains, sediment, algae, animals (invertebrates and vertebrates), plankton, and probably even the plastisphere. Unlike "marine-derived fungi", many marine fungi s.s. have evolved specialized adaptive strategies to ensure successful spore dispersal in the marine environment. Ascomata of arenicolous marine fungi lack long central necks (that would be abraded by the constant movement of the sand grains) and associated ascospores are appendaged and often found to be trapped in sea foam. Submerged wood substrata tend to favor the growth of members of the Halosphaeriaceae, Torpedosporales, and Lulworthiales (taxonomic groups in the fungal kingdom composed predominantly of marine fungi), especially in open oceans, where species are typically characterized by passive ascospore release (deliquescing asci) and appendaged ascospores that aid in floatation and attachment. When occurring on submerged wood, ascomata are found partly or completely immersed within the substrata. Marine Ascomycota are adapted to life in the intertidal zone, and often release spores through the ostiole during periods of low tide, which rest upon the ascomata and become washed off into the water column during high tide, and their ascospores often have gelatinous sheaths (Dothideomycetes), appendages (Sordariomycetes), or, in some species, both to aid in the attachment of spores to substrata. Many conidia (asexual spores) of marine fungi are branched or ornamented or extremely long in size to aid in dispersal and floatation, many of which can be found trapped in seafoam along the shoreline.

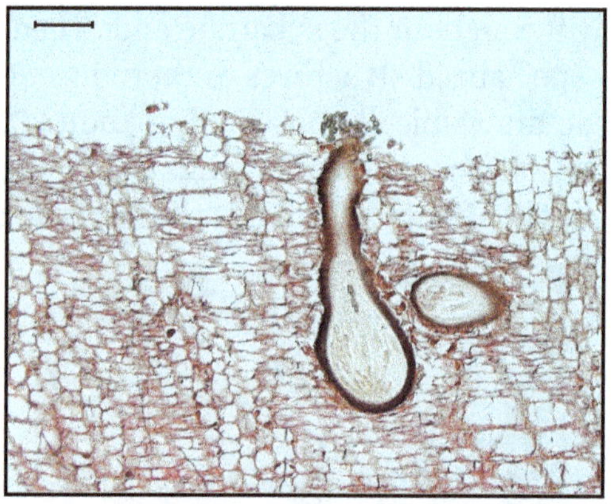

A cross-section micrograph of driftwood-embedded Dyfrolomyces marinospora ascomata (fruiting bodies) releasing ascospores (scale bar = 100 μm)

Marine fungi abound in intertidal zones, such as mangrove ecosystems, due to the associated and abundant organic detritus; over 280 species of fungi have been described from submerged mangrove substrata. Algae dominate marine habitats in temperate regions and approximately 80 fungal species have been associated with algae either as parasites or symbiosis; however, algae as a source of marine fungal isolates have been grossly neglected and further isolation efforts will likely yield undiscovered species diversity. Estuarine (or salt) marshes occur in the intertidal zone and are dominated by the growth of salt-tolerant plants. Few species of salt-tolerant plants have been surveyed; Juncus roemerianus and Spartina species are an exception (due to their dominance in this ecosystem) and are associated with a rich diversity of fungal species.

Although marine fungi in temperate and tropical areas are the most studied, a rich and diverse mycota exist in polar areas. Fungi in these cold environments are associated with similar substrate relationships as in warmer regions. Sea-ice, however, is a habitat unique to polar marine region and supports a diverse community of microorganisms, including fungi. Members of Chytridiomycota and Dikarya (Ascomycota and Basidiomycota) can be found dwelling in high abundance in sea-ice. Studies targeting fungi in macroalgae in Antarctica have revealed high fungal richness. In the Arctic, 100 species of marine fungi have been morphologically detected, whereas molecular data indicates the presence of thousands of fungi in the Arctic marine environment.

Natural Products from Marine Fungi (Sensu Stricto)

The phylogenetic breadth of marine fungi s. within the fungal kingdom is extensive, including marine-specific yeasts, chytrids, and other basal fungal groups; however, fungi in the Dikarya (ascomycetes and basidiomycetes) and the Zoopagomycotina are richer in the number and diversity of secondary metabolite gene clusters and therefore are those most likely to yield the greatest diversity of natural products. Filamentous marine fungi (ascomycetes in particular) have proven to be a plentiful source of new natural products. For example, from an isolate of Aigialus parvus (a marine ascomycete occurring on submerged wood), investigators obtained a total of 11 new natural products representing several different structural classes (aigialomycins A–G, aigialospirol and associated derivatives, and aigialone). Halorosellinia oceanica (syn. Hypoxylon oceanicum), a marine ascomycete associated with intertidal wood in mangrove habitats, studied by Wyeth for associated antifungal properties, led to the discovery of three new lipodepsipeptides (15G256ε, 15G256γ, 15G256δ), five new macrocyclic polyethers (15G256α, 15G256β, 15G256α-1, 15G256ι, and 15G256ω), and five new linear polyesters (15G256α-2, 15G256β-2, 15G256ο, 15G256ν, and 15G256π). New natural products obtained from marine fungi are associated with a range of biological activities (representative structures are presented in figure). For example, an isolate of the obligate marine species, Zopfiella marina, isolated from marine mud (obtained from a depth of 200 m) was found to produce the potent antifungal metabolite, zofimarin, a sordarin derivative. Aigialomycin D demonstrated antiplasmodial activity and cytotoxicity against human cell lines (where cytotoxicity has been attributed to cyclin-dependent kinase/glycogen synthase kinase-3 inhibition).

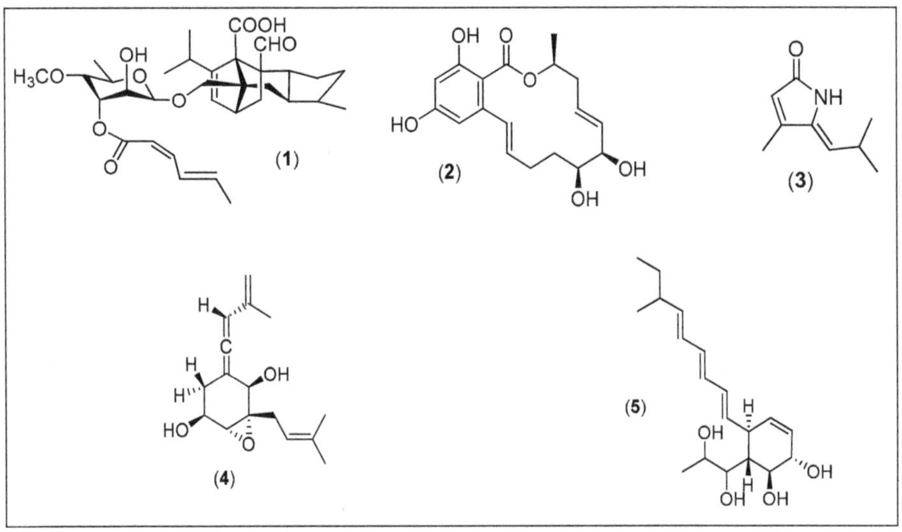

New natural products of marine fungi s: (1) zofimarin, (2) aigialomycin D, (3) pulchellalactam, (4) spartinoxide and (5) spartinol C

The metabolite, pulchellalactam, obtained from a Corollospora pulchella strain isolated from driftwood exhibited inhibitory activity against CD45, an essential transmembrane protein tyrosine phosphatase associated with T- and B-cell antigen receptor signaling. Discovery work involving another marine fungus, Phaeosphaeria spartinae (=Leptosphaeria spartinae), isolated from the inner tissues of the marine alga, Ceramium sp., resulted in the characterization of eight new natural products, including spartinoxide and spartinol C, demonstrating inhibition of the enzyme human leukocyte elastase; a disease target associated with pulmonary emphysema, rheumatoid arthritis, and cystic fibrosis. These examples emphasize that marine fungi can be isolated from a range of different substrata and can produce unique and previously undiscovered natural products. The following text will provide information regarding several methods that are used by mycologists (and some natural product chemists) to isolate truly unique isolates of marine fungi s.

References

- Marine-microorganism: imedpub.com, Retrieved 21 February, 2019
- Marine-Microbes-in-Different-Fields-of-Ecology: researchgate.net, Retrieved 10 May, 2019
- Marine-viruses, marine-microbes: aims.gov.au, Retrieved 23 March, 2019
- Marine-Viruses: researchgate.net, Retrieved 3 July, 2019
- Marine-bacteria, earth-and-planetary-sciences: sciencedirect.com, Retrieved 11 April, 2019
- Protist-definition-2291741: thoughtco.com, Retrieved 30 January, 2019
- Marine-Microalgae: researchgate.net, Retrieved 4 June, 2019
- Phytoplankton: newworldencyclopedia.org, Retrieved 12 August, 2019

4

Marine Mammals

The aquatic mammals which rely on marine ecosystems such as the ocean for their sustenance and existence are termed as marine mammals. There are a number of categories of marine mammals such as Cetaceans, Pinnipeds, Mustelidae and Sirenians. This chapter has been carefully written to provide an easy understanding of the varied facets of these types of marine mammals.

Marine mammals are mammals that have adapted to life in the ocean. They some of the most notable figureheads of the marine community Marine mammals are found in marine ecosystems around the globe. They are a diverse group of mammals with unique physical adaptations that allow them to thrive in the marine environment with extreme temperatures, depths, pressure, and darkness. Marine mammals are classified into four different taxonomic groups: cetaceans (whales, dolphins, and porpoises), pinnipeds (seals, sea lions, and walruses), sirenians (manatees and dugongs), and marine fissipeds (polar bears and sea otters).

Marine mammals evolved from their land dwelling ancestors over time by developing adaptations to life in the water. To aid swimming, the body has become streamlined and the number of body projections has been reduced. The ears have shrunk to small holes in size and shape. Mammary glands and sex organs are not part of the external physiology, and posterior (hind) limbs are no longer present.

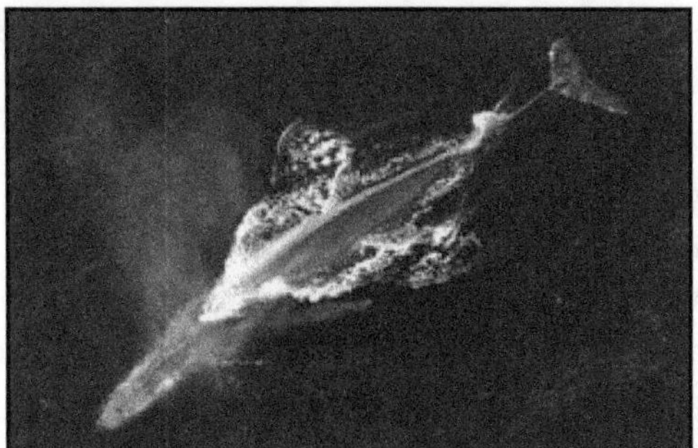

Mechanisms to prevent heat loss have also been developed. The cylindrical body shape with small appendages reduces the surface area to volume ratio of the body, which reduces heat loss. Marine mammals also have a counter current heat exchange mechanism created by convergent evolution where the heat from the arteries is transferred to the veins as they pass each other before getting to extremities, thus reducing heat loss. Some marine mammals also have a thick layer of fur with a water repellent undercoat and a thick layer of blubber that can't be compressed. The blubber provides insulation, a food reserve, and aids with buoyancy. These heat loss adaptations can also lead to overheating for animals that spend time out of the water. To prevent overheating, seals or sea lions will swim close to the surface with their front flippers waving in the air. They also flick sand onto themselves to keep the sun from directly hitting their skin. Blood vessels can also be expanded to act as a sort of radiator.

One of the major behavioral adaptations of marine mammals is their ability to swim and dive. Pinnipeds swim by paddling their flippers while sirenians and cetaceans move their tails or flukes up and down.

Some marine mammals can swim at relatively high speeds. Sea lions swim up to 35 kph and orcas can reach 50 kph. The fastest marine mammal, however, is the common dolphin, which reaches speeds up to 64 kph. While swimming, these animals take very quick breaths. For example, fin whales can empty and refill their huge lungs in less than 2 seconds. During dives marine mammals' larynx and esophagus close automatically when they open their mouths to catch prey. Oxygen is stored in hemoglobin in the blood and in myoglobin in the muscles. The lungs are also collapsible so that air is pushed into the windpipe preventing excess nitrogen from being absorbed into the tissues. Decreasing pressure can cause excess nitrogen to expand in the tissues as animals ascend to shallower depths, which can lead to decompression sickness aka "the bends." Bradycardia, the reduction of heart rate by 10 to 20%, also takes place to aid with slowing respiration during dives and the blood flow to non-essential body parts. These adaptations allow sea otters to stay submerged for 4 to 5 minutes and dive to depths up to 55 m. Pinnipeds can often stay down for 30 minutes and reach average depths of 150-250 m. One marine mammal with exceptional diving skills is the Weddell seal, which can stay submerged for at least 73 minutes at a time at depths up to 600 m. The length and depth of whale dives depends on the species. Baleen whales feed on plankton near the surface of the water and have no need to dive deeply so they are rarely seen diving deeper than 100 m. Toothed whales seek larger prey at deeper depths and some can stay down for hours at depths of up to 2,250 m.

Marine mammals are often very social animals. Dolphins travel in pods (schools) and catch rides on the bow waves of boats. Marine mammals are also known to help each other when one member of the group is injured. There have been accounts of members of a pod refusing to leave the wounded or dying, a trait often exploited by whalers. Cetaceans (whales and dolphins) often hunt together, often with one leading the pod to act as a scout when entering unfamiliar territory. This close knit socialization is thought to be a factor in some whale stranding when a pod follows one or more members of the group that have become disoriented due to storm, illness, or injury.

Many marine mammals also participate in yearly migrations, either in groups or individually. Toothed whales are an exception and only move about in search of food, but some baleen whales (such as gray whales) embark on extremely long migrations, moving from tropical breeding grounds in winter to feeding areas in colder waters during the summer.

Communication

Marine mammals are capable of sophisticated communication because they live in a world dominated by sound, which travels much more efficiently through water than through air. Dolphins communicate with sound to coordinate hunts; humpback whales sing to attract females. Female pinnipeds and their pups recognize each other by their "voices." Slapping the surface during breaching can be heard for miles. Whales have no vocal cords; they warble for up to 30 minutes between breaths just by recycling air. They also emit low frequency sounds that can be heard by humans such as grunts, barks, squeaks, chirps, or even moos. These noises are thought to be associated with different moods and are believed to be used as social or sexual cues during communication. They might also serve as a signature to allow one animal to be recognized by another. Certain pods are known to even have dialects than can be distinguished from others.

Echolocation is a skill that only toothed cetaceans, bats and a few birds have perfected. They send out rapid sound pulses and listen to their echo to find prey and determine their surroundings. It is thought that sperm whales also use echolocation to stun squid with loud clicks. Clicks can be repeated at different frequencies with low frequencies traveling long distances that are highly penetrating. Toothed whales have a structure called the melon on their forehead that focuses and directs the sound waves; incoming sounds are received primarily in the lower jaw, which is filled with fat or oil that transmits the sound to the inner ear.

Conservation Status

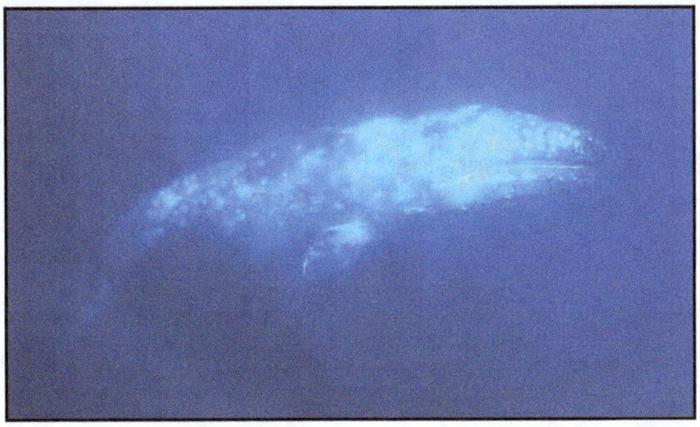

Marine mammals are greatly influenced by their interactions with humans, either directly or indirectly. Fishing takes the lives of at least hundreds of whales, dolphins, and seals every year that drown when they become tangled in fishing nets. Drift nets meant for fish catch anything that goes by, including dolphins, sharks, sea turtles, seals, seabirds, and other marine life. Pollutants also collect (by a process called bio magnification) in the blubber of these marine mammals and many are also hunted by humans. Seals and sea otters are still sought after for their pelts in some places. Sirenians are exploited for their meat, which apparently tastes like veal, and for their skin and oil-rich blubber.

Whales have had one of the biggest population declines due to human hunting. Whaling can be traced back to Native Americans who hunted grey whales, but large scale whaling didn't occur until the 1600's when Europeans began whaling followed soon after by Americans, who dominated worldwide whaling. Large-scale whaling started from open boats where harpoons were used. The blubber was used to make soap and lamp oil and baleen was used as stays for corsets and other garments. Meat and other valuable whale parts were also sold. In the 1800's fast steamships began to be used for whaling and even the faster whales like the blue and fin whales couldn't escape the devastating harpoons. The low reproductive potential of these whales caused the whale global populations to be drastically reduced. The first species to be depleted was the right whale, so named because they were the "right" whale to be killed since they floated after being killed. Whaling nations even developed factory ships to process whole carcasses. Blue whales were particularly sought after because there was so much oil, about 9,000 gallons, in an average whale. As the blue and right whale populations began to dwindle, whalers moved to the fin whale and then the smaller see whale. The disappearance of commercially profitable whales forced the whalers to move to smaller, less profitable whales.

Marine Mammal Adaptations

Deep Diving

Generally, marine mammal lungs are proportionately smaller than humans', but they:

- Use oxygen more efficiently. They fill their lungs and exchange 90% of their air in each breath, have high blood volume, and their blood chemistry allows greater oxygen retention (the high red blood cell count and increased myoglobin make their muscle tissue and blood dark red).
- Have a high tolerance to lactic acid and carbon dioxide. Their muscles can work anaerobically (without oxygen) while they hold their breath.
- Can tolerate tremendous atmospheric pressure at great depths. Lungs and ribs are collapsible, air spaces are minimized, and nitrogen absorption is limited.

Swimming Adaptations

- Drag is reduced by hydrodynamic body forms.
- Appendages are modified for maximal propulsion and minimal drag.

Thermoregulation

- A large body with small surface-to-volume ratio reduces heat loss. Blubber or thick underfur is used as insulation.
- Complex circulatory system in extremities is used to conserve and dissipate heat.
- Young pinnipeds and cetaceans grow fast on milk with 40–50% fat (human milk is 3.3% fat).

Water Conservation

Most marine mammals rarely drink fresh water; instead, they:

- Utilize water present in their food, inspired air, and blubber.
- Have specialized kidneys that produce urine that is saltier than seawater.

Sensory Adaptations

Marine mammals communicate underwater with sound, and many species use sound (echolocation) to locate prey. Tactile senses are acute. Pinnipeds and fissipeds have well-developed facial whiskers.

Cetacean

Cetacean (order Cetacea) is any member of an entirely aquatic group of mammals commonly known as whales, dolphins, and porpoises. The ancient Greeks recognized that cetaceans breathe air, give birth to live young, produce milk, and have hair—all features of mammals. Because of their body form, however, cetaceans were commonly grouped with the fishes. Cetaceans are entirely carnivorous, although members of the order Sirenia (manatees, dugongs, and Steller's sea cow) were once referred to as the "herbivorous Cetacea." In the past cetaceans were important resources, but by the end of the 20th century their economic importance was almost solely due to whale watching, a tourist activity and major source of income for certain coastal regions of many countries.

Form And Function

Features

Cetaceans were commonly grouped with the fishes like whales, dolphins, and porpoises. The features and functions of Cetacean are following i.e. given below:

Body Surface

The hair covering that is common to mammals is drastically reduced in cetaceans, likely because hair is a poor insulator when wet and increases drag during swimming. Hairs on cetaceans are restricted to the head, with isolated follicles occurring on the lower jaw and the snout. These are thought to be remnants of sensory whiskers (vibrissae). External pigmentation is important to many animals as a basis for individual recognition and species recognition. Hair defines the colour pattern of most mammals, but, because cetaceans have very little hair, the outer layer of skin (epidermis) produces their markings, most commonly in shades of black and white. The appearance of some cetaceans is affected by various organisms living on or in the skin. Examples include yellow algae that colour the lower body surface of blue whales (Balaenoptera musculus) and the variety of whitish organisms living on bodies of gray whales (Eschrichtius robustus) and right whales (family Balaenidae).

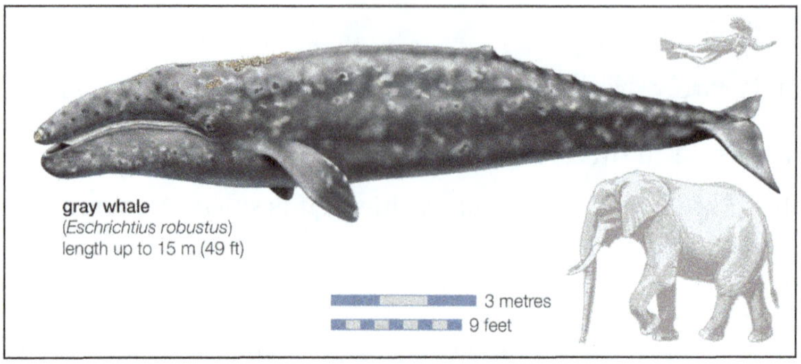

Gray whale (Eschrichtius robustus)

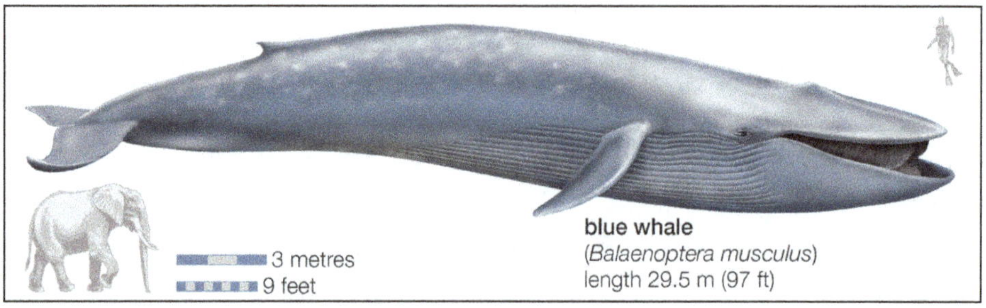

Blue whale (Balaenoptera musculus)

Locomotor Adaptations

The most noticeable adaptation of cetaceans to life in the water is their locomotive system. Because cetaceans descended from mammals that moved their limbs in a vertical plane rather than in a horizontal plane, they use vertical strokes when they swim, instead of horizontal strokes like a crocodile or fish. Cetaceans evolved from four-legged (quadruped) terrestrial animals, for which limbs played a primary role in movements, into virtually limbless aquatic creatures living in an environment where the back muscles are more important. Forelimbs are still present but are reduced to finlike flippers having shortened arm bones and no individual fingers. The hind limbs are lost entirely; only vestigial elements sometimes remain internally. Pelvic remnants occur in all cetacea but the dwarf and pygmy sperm whales. Flippers help to steer, while the back muscles, which are very large, drive the tail to propel the animal. Cetaceans have developed horizontal flukes that increase the propulsion area driven by the back muscles. Like fish, almost all cetaceans possess a dorsal fin that serves as a keel. The dorsal fin and flukes are composed of connective tissue, not bone. Other connective tissue, such as external ears, has been lost, and the male genitalia have moved internally.

Humpback whale (Megaptera novaeangliae) breaching

Respiration

Normally, cetaceans breathe while moving through the water and spend only a short time at the surface, where they exhale in explosive ventilation called a blow. The blow is expelled forcibly and can be compared to a cough. Cetaceans use up to 80 percent of their lung volume in a single breath, in contrast to humans, who use only 20 percent. The blow is visible because of water condensation and mucous particles; blows of blue whales are frequently more than 6 metres (20 feet) high. When a terrestrial mammal loses consciousness, it breathes reflexively, but breathing is not a reflex in cetaceans. Thus, when a cetacean loses consciousness, it does not breathe and quickly dies. For this reason, veterinarians had to perfect respirators before dolphins could be successfully anesthetized.

Circulation and Thermoregulation

Cetaceans, like all mammals, have a four-chambered heart with paired ventricles and auricles. The pattern of circulation is similar to that of other mammals, with the exception of a series of well-developed reservoirs for oxygenated blood called the rete mirabile, for "marvelous network." These provide bypasses that enable cetaceans to isolate skeletal muscle circulation during diving while using the oxygen stored in the remaining blood to maintain the heart and brain—the two organs that depend on a constant supply of oxygen to survive.

Water conducts heat much more rapidly than air and is colder than the mammalian body temperature of about 37 °C (98.6 °F). Cetacean evolution has countered this problem in three ways: reducing external appendages that lose heat, developing an insulating layer of blubber, and developing counter current circulation to minimize heat loss. The reduction of various appendages as mentioned above also facilitates locomotion in water.

In whales, a layer of the skin (dermis) has evolved into a blanket of blubber, which is extremely rich in fats and oils and therefore conducts heat poorly. This blanket covers the entire body and is up to 30 cm (12 inches) thick in large whales, making up a significant portion of the animal's weight. The oil yield of blubber from a blue whale, for example, was up to 50 tons.

The most important mechanism in cetacean thermoregulation is the development of counter current blood exchange, an adaptation that allows the animal to either conserve or dissipate heat as needed. Blood that drains from the surface of the skin has been cooled by close contact with the external environment, and it can return to the cetacean's heart via two different routes. If it returns by the peripheral route, the blood courses back to the heart through superficial veins, where it continues to lose heat and arrives at the heart cool. This dumps the animal's excess heat to the environment. Such heat shedding is particularly important to large whales because of their enormous surface area-to-volume ratio. If, however, the body temperature of the whale is already cool, the oxygen-depleted venous blood can instead return to the heart through vessels that are wrapped around arteries carrying warm blood to the periphery of the animal. Along this route the venous blood is warmed by the arterial blood and arrives at the heart warm. The arterial blood, having transferred its heat into the venous blood rather than the environment, arrives precooled at the surface of the skin.

Feeding Adaptations

Before cetaceans evolved aquatic adaptations, they had a fully differentiated set of teeth (heterodont dentition), including incisors, canines, premolars, and molars. As the animals became more adapted to aquatic locomotion and lost the ability to manipulate food with their forelimbs, they started grabbing their food and swallowing it whole. In toothed whales (suborder Odontoceti), heterodont dentition declined and was replaced with a homodont dentition in which every tooth is a simple cone. The number of teeth varies among toothed whales, from two in the beaked whales (family Ziphiidae [Hyperoodontidae in some classifications]) to 242 in the La Plata river dolphin (Pontoporia blainvillei), to allow efficient capture of prey. Baleen whales (suborder Mysticeti), on the other hand, have lost all teeth in both jaws and instead have two rows of baleen plates in their upper jaws only. This apparatus enables baleen whales to consume vast quantities of small prey in a single mouthful.

Killer whale (Orcinus orca)

In general, whales have relatively large mouths. The mouth of one adult bowhead, or Greenland right whale (Balaena mysticetus), measures five metres long and three metres wide and is the biggest oral cavity on record. The stomach in cetaceans is composed of four compartments: forestomach, main stomach, connecting chambers, and pyloric stomach. The forestomach is actually a dilation of the esophagus and is lined with simple epithelium (layers of flattened cells). It acts merely as a holding chamber and therefore is not a true stomach. The main stomach, lined with active gastric epithelium, is the first true digestive compartment, and it is followed by the small connecting chambers and the pyloric stomach. From there, food enters the small intestine through the pyloric sphincter and the duodenal ampulla. Most cetaceans do not have a cecum or appendix, and in most there is no anatomic difference between the small and the large intestine.

Greenland right whale, or bowhead (Balaena mysticetus)

Senses

The sensory system of any animal can be divided into somesthetic senses—those relating to the whole body—and special senses associated with particular organs such as the eyes and ears. Somesthetic senses are broken down into exteroceptive (initiated by stimuli outside the body), proprioceptive (initiated within the body, determining the orientation of body parts relative to one another and the orientation of the body in space), and visceral (usually from internal organs and usually painful). Cetaceans, as far as is known, are subject to the familiar exteroceptive sensations. For example, captive and stranded animals respond to stimuli of touch, pain, and heat. Because precise assessment of the other somesthetic modalities (proprioceptive and visceral) is difficult, scientists have simply assumed their presence.

The special senses respond to stimuli registered by specialized organs or tissues. One way to quantify the presence of a special sense in an animal is to consider the organs involved.

Smell

The sense of smell can be defined as those sensations carried from nose to brain by the olfactory nerve. Toothed whales have lost the olfactory nerve, so by definition they are incapable of smelling. On the other hand, they do use "quasi-olfaction". Baleen whales have retained this nerve and have a reduced area for olfaction in the nasal passage, but this sense is active only while the animal is breathing at the surface.

Taste

Captive dolphins (family Delphinidae) commonly exercise food taste discrimination that is comparable to the human ability, in spite of the fact that the presence of taste buds in cetaceans has not been demonstrated. Regardless, dolphins have been shown to be sensitive to the standard four qualities of taste: sweet, salty, sour, and bitter. It has been established that the bottlenose dolphin (Tursiops truncatus) has a highly effective sense, called quasi-olfaction, operating through pits in the back of the tongue. This sense permits dolphins to experience what would be classified as smell, but quasi-olfaction does not involve the nasal passages.

Sight

Cetaceans have well-developed eyes and good vision. The popular notion that whales have reduced vision is probably based on the relative size of their eyes, but this assumption is functionally incorrect. Vision in both the water and the air has been experimentally evaluated in captive dolphins and found to be excellent. They have binocular vision over at least part of the visual field but are largely insensitive to colour. In one genus of river dolphin (Platanista of the muddy Ganges and Indus rivers), the eyes are reduced to organs that can detect only the difference between light and dark. The external opening for the eye is a slit only 2–3 cm (about an inch) long.

Whales and dolphins have long been known to possess an acute sense of hearing. When approaching whales, whalers muffled their oars to prevent the animals from hearing them. Research done with captive animals in the 1950s quantitatively demonstrated that dolphins both produce and are sensitive to sounds into the ultrasonic range. Dolphins and porpoises were found to have the ability to derive information about their environment by listening to echoes of sounds that they have produced (echolocation). The amount of information obtained by an echolocating dolphin is similar to that obtained with the eyes of a sighted human.

The sound sensitivity of dolphins falls off near the bottom of the human acoustic spectrum (40–50 hertz), but this is the beginning of the range used by the large baleen whales. Fin whales (Balaenoptera physalus) and blue whales have been recorded producing subsonic sounds around 10 hertz and are capable of producing extremely loud noises at those frequencies. The strength of these vocalizations enabled one blue whale to be followed by fixed hydrophone arrays on the ocean bottom for 43 days over a course of 2,700 km (1,700 miles).

Magnetic Sensitivity

Much interest has been shown in various animals' ability to sense the Earth's magnetic field. It has been demonstrated that birds and fish use magnetoreception in migration, and theories to explain why cetaceans beach themselves in mass stranding's have included magnetic detection. Although magnetite has been found in some skulls of the common dolphin (Delphinus delphis), it has not been found in other specimens of the same species, and no conclusive data indicate its biological use.

Types of Cetaceans

Cetaceans are divided into two major suborders: Mysticeti (baleen whales) and Odontoceti (toothed whales, including whales, dolphins, and porpoises).

- Mysticeti: The baleen whales (Mysticeti) are characterized by the baleen, a sieve-like structure in the upper jaw made of the tough, structural protein keratin. The baleen is used to filter plankton from the water. The mysticete skull has a bony, large, broad, and flat upper jaw that is placed back under the eye region. They are characterized by two blowholes. Baleen whales are the largest whales. The families of baleen whales include the Balaenopteridae (humpback whales, fin whales, Sei Whale, and others), the Balaenidae (right and bowhead whales), the Eschrichtiidae (gray whale), and the Neobalaenidae (pygmy right whales), among others. The Balaenopteridae family (rorquals) also includes the Blue Whale, the world's largest animal.
- Odontoceti: The toothed whales (Odontoceti) have teeth and prey on fish, squid, or both. This suborder includes dolphins and porpoises as well as whales. In contrast with the mysticete skull, the main bones of the odontocete upper jaw thrust upward and back over the eye sockets. Toothed whales have only one blowhole. An outstanding ability of this group is to sense their surrounding environment through echolocation. In addition to numerous species of dolphins and porpoises, this suborder includes the Beluga whale and the sperm whale, which may be the largest toothed animal to ever inhabit Earth. Families of toothed whales include, among others, the Monodontidae (belugas, narwhals), Kogiidae (Pygmy and dwarf sperm whales), Physteridae (sperm whale), and Ziphidae (beaked whales).

The terms whale, dolphin, and porpoise are used inconsistently and often create confusion. Members of Mysticeti are all considered whales. However, distinguishing whales, dolphins, and porpoises among the Odontoceti is difficult. Body size is useful, but not a definitive distinction, with those cetaceans greater than 9ft (2.8m) generally called whales; however, some "whales" are not that large and some dolphins can grow larger. Scientifically, the term porpoise should be reserved for members of the family Phocoenidae, but historically has been often applied in common venacular to any small cetacean. There is no strict definition of the term dolphin.

Distribution and Migration

Cetaceans are distributed in all the world's oceans from the far polar reaches to the Equator. They concentrate in areas of increased biological productivity, such as upwellings where there is an abundant supply of food. Some species are coastal and some are pelagic, dwelling farther offshore. The centres of the ocean basins appear not to have any concentrations of whales or dolphins. Some small cetaceans are distributed in major river systems, particularly river dolphins of the family Platanistidae. Members of this family are found in the Amazon, Orinoco, La Plata, Yangtze, Ganges, and Indus rivers and surrounding drainage waters. Members of other families, particularly the Delphinidae and Phocoenidae, spend part of their time in fresh water.

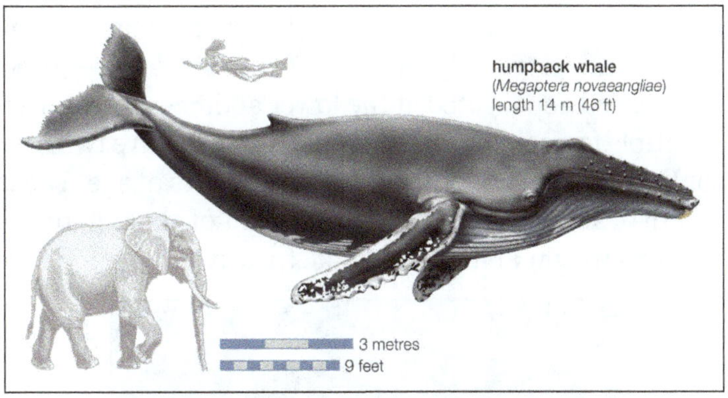

Humpback whale (Megaptera novaeangliae)

As a rule, large whales have north and south seasonal migrations, spending summers in high latitudes near the poles, where there is an abundant food supply, and moving toward the Equator in the fall to breed. Some populations of these species, however, reside in one locality all year. One of the greatest migrations is undertaken by the California population of the gray whale, which summers in the Bering and Chukchi seas of the Arctic and winters in lagoons off the coast of Baja California—a journey of 5,000 km (about 3,000 miles) each way.

Migratory whales usually do not cross the Equator, and that has led to the development of genetically separate populations in the north and South Ocean basins. However, one humpback whale (Megaptera novaeangliae) was photographically identified near the Antarctic Peninsula and was later sighted on the coast of Colombia, having covered at least 8,334 km in both the South Atlantic and North Atlantic Ocean basins. Some sperm whales (Physeter catodon) sexually segregate on their migrations, with larger and older males going much farther in a polar direction during the summer.

Diet

All cetaceans are carnivores and do not consume plants or algae as food. The large baleen whales eat schooling organisms that range in length from minute drifting mollusks, copepods (1 cm or less), krill (1–5 cm), and small fish and squid up to about 40 cm. All these are consumed by whales in vast quantities with each concentrated mouthful. Certain smaller baleen whales, such as the minke whale (Balaenoptera acutorostrata), also pursue individual fish up to 1 metre long. Toothed whales, which range in length from about 1 metre for the finless porpoises (Neophocoena phocaenoides and N. asiaeorientalis) to 20 metres for the sperm whale, eat an enormous variety of prey ranging from small shrimp, fish, and squid to bluefin tuna (3 metres long) and giant squid.

Cetaceans void their solid digestive waste products as pastelike feces, enabling the retention of intestinal water. Liquids are excreted in the urine. Biologists do not entirely understand the function of ambergris, which seems to be a normal digestive secretion of the sperm whale.

Reproduction

Cetacean breeding is seasonal, usually in the winter, and females normally calve once every two years. As mammals, they reproduce by internal fertilization. The testes and penis of the male are internal, but the penis is capable of being extended and introduced to the female during mating. After the female's egg has been fertilized, she carries the fetus for about a year, although some toothed whales have gestations of up to 18 months. Cetaceans give birth tail first, opposite of most terrestrial mammals. The mothers produce extremely rich milk for their young—50 percent fat is common. The mammary glands are paired and located at the lower abdomen, just forward of the anal-genital slits. One calf is born (multiple fetuses have been found, but never live twins), which is weaned at six months to a year but continues to grow rapidly until 5 to 10 years of age. The largest whale, the blue whale, is born with a length of about 7.3 metres and a weight of about 3 tons; it grows an average of 0.3 metre per week and gains weight at a rate of 90 kg (nearly 200 pounds) per day.

Bottlenose dolphin: A newborn bottlenose dolphin calf swimming alongside its mother at SeaWorld

Age and Growth

Generally, male cetaceans reach a slightly greater size than females, but there are many exceptions; female baleen whales and some of the beaked whales, river dolphins, and porpoises tend to be slightly larger than males of their species. Male sperm whales, on the other hand, are on average 50 percent larger than females. Physical maturity is defined specifically in cetaceans as the point when all of the vertebrae stop growing, which occurs at an age of about 8–25 years. Historical estimates of whale longevity are limited because of difficulty measuring age in older whales. Modern techniques have determined some fin whales to be 100 years old, some humpbacks 96, and some blue whales 90. But the longest-lived cetacean by far is the bowhead, a right whale that can survive for more than 200 years.

Sexual maturity occurs at an age of about 6–10 years and is defined in cetaceans as the age at which the females start ovulating and are capable of becoming pregnant. In most species, females continue to ovulate throughout their lives and thus can produce young until death. However, females in several toothed whale species—including killer whales, false killer whales, short-finned pilot whales, belugas, and narwhals—experience menopause (in which ovulation stops), and the rest of their lives are spent in a post-reproductive state. Authorities speculate that this adaptation may have evolved in order for older females to assist with the rearing of young.

Abundance

Counting animals that can be spread over wide areas of the world's oceans and are visible for only a few seconds while they breathe is extremely difficult and expensive. The largest populations existed in the oceans around Antarctica, where harsh and remote conditions make biological research difficult and infrequent. The abundance of cetaceans is thus hard to estimate accurately, but whale populations have varied over the years, depending largely upon human activities.

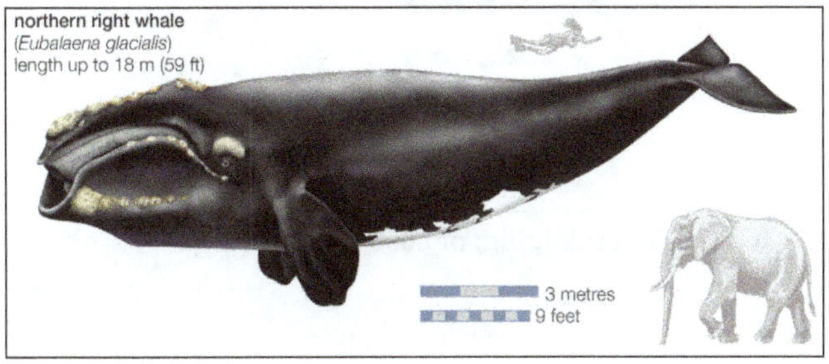

Northern right whale (Eubalaena glacialis)

Biologists estimate that there were 228,000 blue whales and 548,000 fin whales in the world's oceans when modern whaling began in the early 20th century. At the beginning of the 21st century, there were an estimated 14,000 blue whales and 120,000 fin whales left. California gray whales were thought to number 20,000 in 1847, then were hunted until they were thought to be extinct in the 1920s. Since then the species has recovered under protective legislation, and its population has been estimated to be more than 26,000.

Japanese factory ship hauling a minke whale through a slipway in the ship's stern

Until the early 21st century, the only cetacean population to be completely exterminated was the Atlantic gray whale, which was gone in the early 1700s; however, the baiji, or Chinese river dolphin (Lipotes vexillifer), a species restricted to the Yangtze River (Chang Jiang), was widely believed to be extinct. In addition, some smaller cetacean species with limited distributions—such as the Gulf of California porpoise (Phocoena sinus), Indus susu (Platanista minor), and Ganges susu (Platanista gangetica)—could be in immediate danger of becoming extinct. (Some sources consider P. minor and P. gangetica as subspecies rather than separate species.) Furthermore, many other cetacean species, such as the gray whale and the northern right whale (Eubalaena glacialis), continue to be menaced by ship collisions, pollution, entanglement in commercial fishing equipment, and illegal hunting.

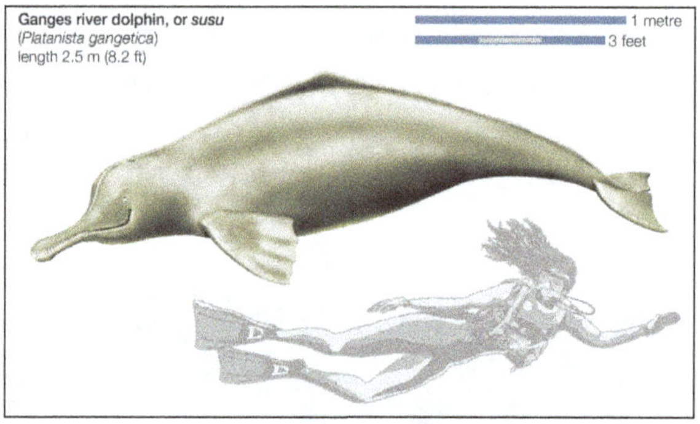

Ganges river dolphin, or susu (Platanista gangetica)

Diseases and Parasitism

Cetaceans can suffer from many of the same diseases and parasites that afflict humans and other mammals: cancer, arthritis, pneumonia, lungworms, tapeworms, and roundworms, to name just a few. In the 1980s various dolphin species experienced epidemics of a morbillivirus, a disease similar to distemper and measles.

Of the parasites and commensal organisms, some are also found on fish and marine turtles, and others are specific to cetaceans. Commensal barnacles are most visible on humpbacks and gray whales, although they occur to a lesser extent on many other baleen and toothed whales. Xenobalanus globicipitis, a unique type of small pseudo-stalked barnacle, occurs on the appendages of cetaceans, including the common bottlenose dolphin. Stalked barnacles can also occur on exposed teeth and can be particularly striking on the tusks of beaked whales.

Minke whale (Balaenoptera acutorostrata)

Different types of commensal or parasitic crustaceans inhabit whales. There is a small commensal copepod, Balaenophilus, that eats the algae on the baleen of some rorqual species. A specific family of amphipods (Cyamidae) called whale lice routinely infest right, humpback, and gray whales and also occur opportunistically on most species of baleen and toothed whales, particularly around wounds. They appear to eat sloughed skin. Another crustacean is clearly parasitic and is a member of the caligoid copepod genus Pennella. It is commonly about 2–10 cm long and lives with its body buried in the blubber or skin of cetaceans and fishes.

Small (4–7 cm) circular scars left in the skin of whales, dolphins, and fish were a mystery during the early 20th century. Explorers reported that the scars were created by an unknown organism that they called the "DWB" or "demon whale biter," which cleanly removed hemispheric chunks of blubber as though extracting them with a razor-sharp scoop. The creature responsible was finally identified in the 1950s as a grazing predator, the cookie-cutter, or cigar, shark (genus Isistius).

Locomotion

Swimming

Cetaceans swim by using vertical tail movements that drive the horizontal flukes up and down, powered by the long epaxial and hypaxial muscles that lie along the spine. The tail flexes through a point between the dorsal fin and the anus, while the thorax and abdomen are relatively inflexible. The body itself acts like a spring to propel the animal through the water with minimal energy.

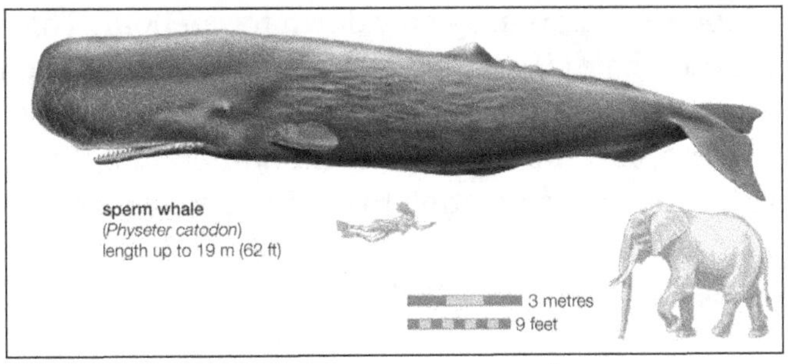

Sperm whale (Physeter catodon)

Much was written about the speeds of cetaceans in the mid-20th century. It seemed that cetaceans could exceed the speed at which turbulence would make locomotion energetically very expensive. However, the swimming-speed figures were estimates that turned out to be very high. Further investigation found that, regardless of size, the cruising speed of most cetaceans is about 2 metres per second (about 7 km, or 4 miles, per hour). A combination of biomechanical and hydrodynamic factors makes this an efficient speed at which to travel. Maximum speeds, however, vary greatly between species.

Common dolphins (genus Delphinus) have been observed keeping pace with boats for a considerable period of time at 36 km/hr (kilometres per hour). Researchers trained Pacific bottlenose dolphins (genus Tursiops) to swim in an open-water environment, thus removing the spatial limitations of a pool while conserving experimental controls. They found that the dolphins could sprint at 29.9 km/hr for 7.5 seconds and could maintain speed at 21.9 km/hr for 50 seconds. When dolphins ride a bow wave, they coast at the speed of the ship while expending very little energy. Fin and blue whales can swim fast enough that a boat must travel in excess of 30 km/hr to catch up to them, and they can maintain speeds of 33–37 km/hr for periods of up to 10–15 minutes. Sonar records indicate that fin whales can sprint at 48 km/hr. Right, humpback, and gray whales, however, can seldom swim faster than 9 km/hr. Sperm whales can cruise at 7.5 km/hr and swim up to 36 km/hr in spurts. The fastest cetacean appears to be the sei whale (Balaenoptera borealis), recorded moving at speeds up to 65 km/hr along the ocean surface.

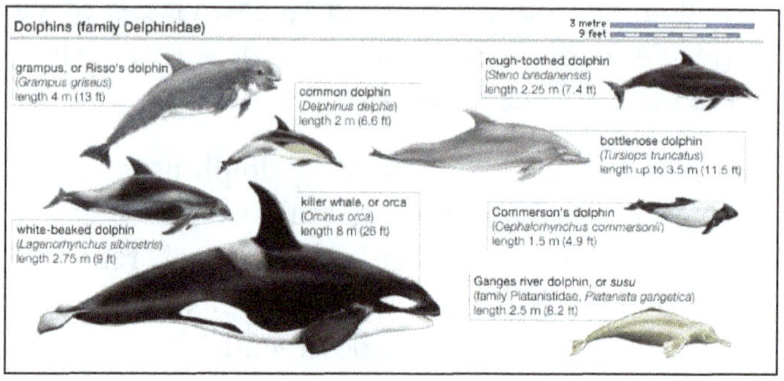

Dolphins (family Delphinidae) and river dolphin (family Platanistidae)

Breathing and Diving

Cetaceans surface periodically to breathe, and the intervals between breaths vary depending on what the animal is doing. Intervals may range from about 20 seconds for dolphins that are actively swimming to 5–10 minutes for a resting blue whale. A common breathing pattern in large whales is to breathe every 20 seconds for 8–10 breaths and then dive for about 10–15 minutes. Most whales stay in the upper 100 metres of water. Deep-diving whales—such as the sperm whale, which has been recorded diving to depths of 1 km—may stay down for an hour. The longest recorded dive is that of a harpooned bottlenose whale (Hyperoodon ampullatus) that dived for two hours, surfaced, and then dived again. Patterns of locomotion and breathing are very important to whale watchers identifying whales at a distance, as different species show different blow heights and shapes. Right whales, for instance, have an unequal inclination to their two nasal passages, so their blows appear in pairs. Humpbacks and gray whales have blown that appear low and wide (bushy), and sperm whales have a bushy blow that is angled to the left and forward.

Leaping and Wave Riding

Small cetaceans "porpoise" when they are swimming rapidly; that is, they rise out of the water in a low leap that keeps the head clear of the water for breathing. Spinner dolphins (Stenella longirostris) frequently leap out of the water while spinning on their long axis, hence their common name. Trained porpoises and dolphins can leap straight up as high as six metres. Leaping is very rare in large whales, but some rorquals (genus Balaenoptera) have been photographed jumping clear of the water.

Whales such as humpbacks (Megaptera novaeangliae) communicate by producing low-frequency sound waves. The animals move sufficiently far beneath the ocean surface before vocalizing, which enables their signals to be heard over hundreds of kilometres

Many small cetaceans play around moving boats, where they bow-ride, taking advantage of their ability to bodysurf and essentially enjoying the free ride in the bow wave created by the vessel. They also practice this behaviour around large whales that are swimming fast enough to produce a bow wave.

Behaviour

Social Behaviour

All cetaceans are social to some extent. The minimum group of mother and calf is commonly expanded to a nuclear family or a group of closely related individuals. A group of cetaceans that normally feed and travel together is called by various names: school, herd, pod, or gam. It is often difficult to define or measure, as its members can be spread over kilometres of ocean but still be in contact with one another. Sometimes these schools coalesce into even bigger groups of more than 1,000. Groups of whales can persist for many years, and studies of coastal dolphins have shown long-term association of dolphins with their mothers. Groups (particularly of small toothed whales) frequently associate with other cetacean species. For example, associations between pilot whales and bottlenose dolphins have been observed, as have associations between common dolphins and fin whales.

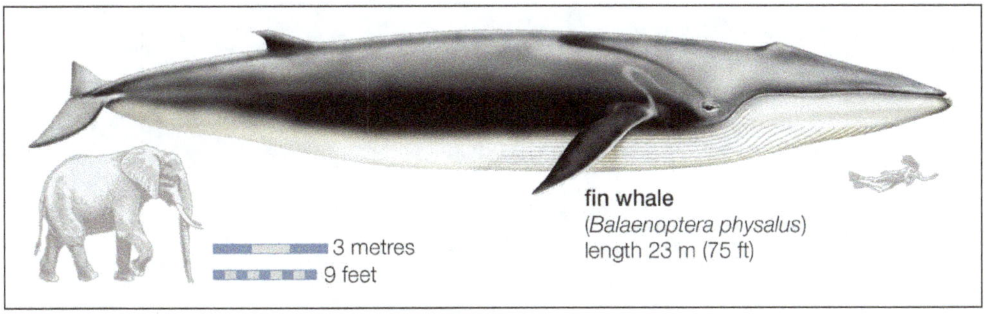

Fin whale (Balaenoptera physalus)

Play is a common behaviour, especially among young animals. Play allows individuals to practice and perfect behaviour patterns, such as aggression, that will be socially useful later in life; a significant portion of play is sexually oriented. Captive dolphins have also been observed playing with fish, birds, and turtles.

Many cetaceans exhibit epimeletic behaviour, in which healthy animals take care of another animal that has become temporarily incapacitated. This is evident when a wounded or sick whale is supported by others or in cases when a dolphin (usually the mother) pushes a dead calf around.

Cetaceans show fright by fleeing from a situation or by bunching up and "milling." The former response has been utilized by fishermen, who drive a whale or school of dolphins into a situation where they can kill it. Milling has been seen in dolphin schools driven into an enclosure or caught in a net; the animals move in a circle or eddying mass, and at the height of this reaction they stop swimming, sink, and die.

Aggression and Defense

Aggression is common among cetaceans and is seen in normal herd behaviour and feeding. One form of aggression helps to establish social hierarchy: the dominant animal nips the less-dominant animal, which produces the tooth scars seen on every adult in the dolphin family (Delphinidae). Mating behaviour also involves biting, as one of the ways males compete for females is by biting and raking the teeth over another male. Adult male beaked whales (family Ziphiidae) have very densely ossified rostra (beaks) used as weapons in combat for females. Another more dangerous means of aggression is head butting. Cetaceans can ram their heads into other individuals and kill them. This has been seen in captivity and in aggressive behaviours toward other species such as sharks and accounts for many of the broken ribs and vertebrae seen in stranded animals.

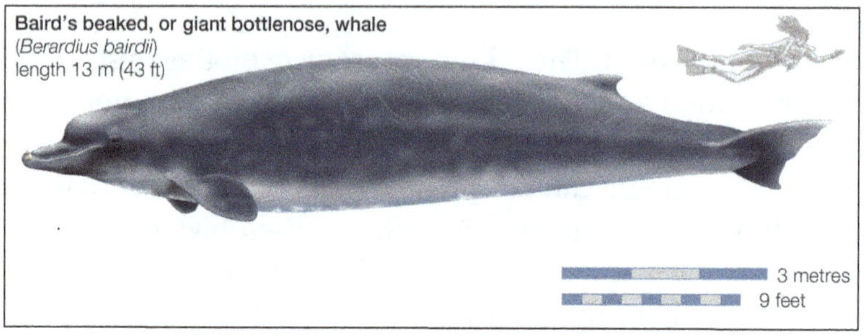

Baird's beaked, or giant bottlenose, whale (Berardius bairdii)

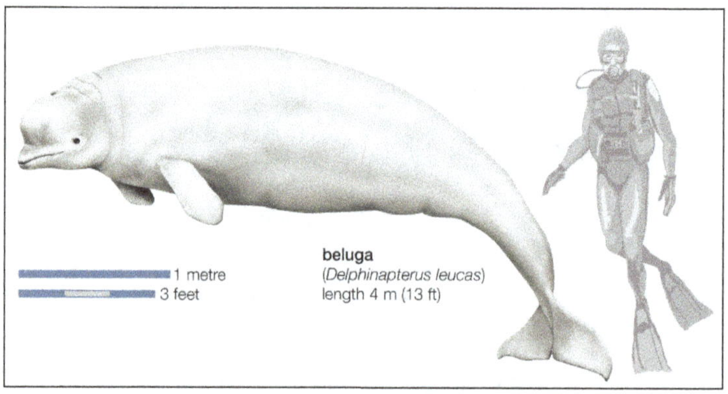

Beluga (Delphinapterus leucas)

Normally, aggression is associated with members of the same species or as a defense response to predation from other species. Although cetaceans can defend themselves by utilizing the behaviours of intraspecific aggression (biting, ramming, and butting), the primary weapon that cetaceans have for self-defense is the tail. Cornered whales slash sideways with their flukes and can incapacitate a bigger whale or a boat. Head butting as a form of defense was immortalized in the 1820 sinking of the whaling ship Essex by a sperm whale.

Courtship and Mating

Sexual behaviour starts early in cetaceans. Young dolphins engage in exploratory sexual behaviour involving their mothers and other members of the school. Self-stimulation is common in both sexes. Male cetaceans perhaps use their penises as a manipulation organ in much the same way that people use their hands. This exploratory behaviour gradually becomes courtship and mating behaviour.

Courtship involves physical and acoustic displays, such as the elaborate songs of male humpbacks, and leads to contact with the flippers and other parts of the body. Successful courtship culminates in mating. Copulation is relatively brief in cetaceans. It can be secretive, or it can be boisterous as in the mating displays of right and gray whales, when a number of males attempt to mate with a single female.

Feeding

Cetaceans hunt as individuals or in schools. When hunting in schools, dolphins or whales herd their prey in order to concentrate a large volume before eating. Hunting alone is preferred where prey is more scattered.

Killer whale (Orcinus orca)

Before they swallow their food, toothed whales disable it; biologists think that some can stun their prey by emitting a high-energy burst of sound. Normally, cetaceans eat animals that can be swallowed intact, as their teeth are shaped for holding, not chewing. If, however, the prey is too large to swallow in one bite, it is ripped into chunks. Killer whales (Orca orcinus) have been seen to grab seals and shake them in the air so hard that the bodies come apart.

Orca, or killer whale (Orcinus orca)

Baleen whales also herd their prey like toothed whales, but they engulf it in either of two feeding methods: gulp or skim. In gulp feeding, the whale opens its mouth to take in a huge mouthful of water, closes its mouth, strains the water out through the baleen apparatus along the sides of the mouth, and swallows its prey. Gulp feeding is common in rorquals, which have ventral grooves that stretch to enlarge the oral cavity. One of the rorquals, the sei whale, as well as the nonrorqual baleen whales (right, bowhead, pygmy right, and gray), skim-feed by locating a concentration of zooplankton prey and swimming through it with the mouth open. Skimming may last up to several minutes until the whales close their mouths to swallow what they have filtered from the water.

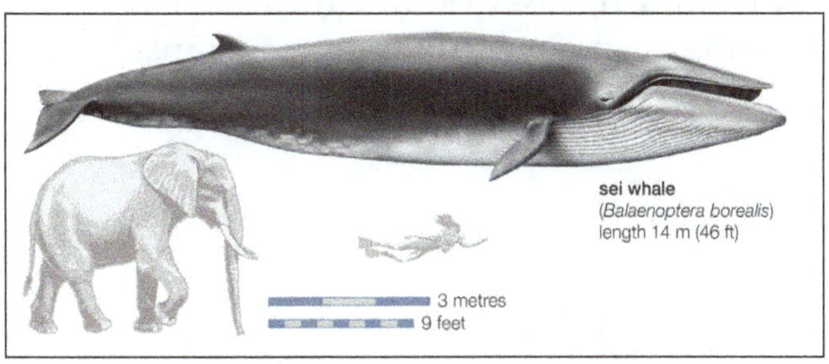

Sei whale (Balaenoptera borealis)

Sleep

Breathing is a conscious activity in cetaceans; they must consciously breathe, or they will drown. Therefore, they cannot enter into what humans understand as unconscious sleep; instead, they have periods of little activity but not total inactivity. Studies of dolphins have revealed that they shut down half of their brain during sleep. The other half of the brain stays awake to signal when to rise to the surface to breathe and to watch for predators and obstacles. Large whales appear to surface-sleep. Floating horizontally just below the water's surface, they move their flukes periodically to rise above the water for a breath.

Intelligence

Although several cetaceans are easily trained and much has been theorized about the possible intelligence of whales and dolphins, little is known for certain. Some researchers equate brain size with intelligence, reasoning that cetaceans should have the capacity for intelligence because they have relatively large brains. The human brain averages about 1.2 kg, the bottlenose dolphin brain about 1.8 kg. The largest cetacean brain recorded was a sperm whale's, weighing 9.2 kg. However, cetaceans may use their increased brain weight for processing acoustic information. In any event, it seems unproductive to compare species with which it is difficult even to communicate until a definition of nonhuman intelligence has been refined.

Stranding

Stranding is a phenomenon that has long fascinated people, and there is fossil evidence of mass strandings from before humans evolved. Many stranded cetaceans are found already dead, and it is not known if they were alive and conscious when they stranded themselves. When a whale or dolphin dies offshore, it usually sinks; if the water is shallow enough to permit decomposition gases to form, it will float ashore, so some stranding represents normal mortality. If infection or some other factor interferes with a cetacean's ability to navigate, it could come ashore while still alive—though most cetaceans have difficulty out of water and usually die. These cases are known (alive or dead) as single strandings. Sometimes up to several hundred toothed whales swim ashore and this phenomenon is known as a mass stranding.

There are no records of the mass stranding of baleen whales; all such events have involved only toothed whales that normally live offshore and may not be familiar with physical borders. Perhaps not realizing that the ocean has a bottom and sides, they may somehow enter shallow water and find themselves unable to deal with the strange environment. Because they are also members of extremely social species and may be kept together by group ties, they may have even greater difficulty extricating themselves. In any case, biologists are beginning to realize that cetaceans are behaviorally complex enough that a simple blanket explanation of mass stranding is not likely to be valid. Biologists have tried to attribute mass stranding to a number of causes:

- Something wrong with the leader of a group
- Epidemic disease
- Getting lost in pursuit of prey

- Parasitic infestation that affects the hearing
- Following migratory routes laid down by remote ancestors
- Magnetic anomalies that lead the school astray
- Behavioral reversion to a period when cetacean ancestors were terrestrial and land was a haven
- Fright reaction to predators
- Failure of echolocation signals to work properly in shallow water
- Overpopulation
- Suicide

Paleontology and Classification

Paleontology

Cetaceans are distant descendants of a group of poorly defined mammals known as condylarths. There is debate as to whether the first cetaceans (archaeocetes) descended from an extinct group of large carnivores called mesonychids or from a group of hoofed herbivores (artiodactyls). The earliest archaeocetes were huge dolphinlike creatures 6 to 10 metres long. Basilosaurus (Zeuglodon) was an unusual genus that was up to 34 metres long, but it apparently gave rise to no descendant groups.

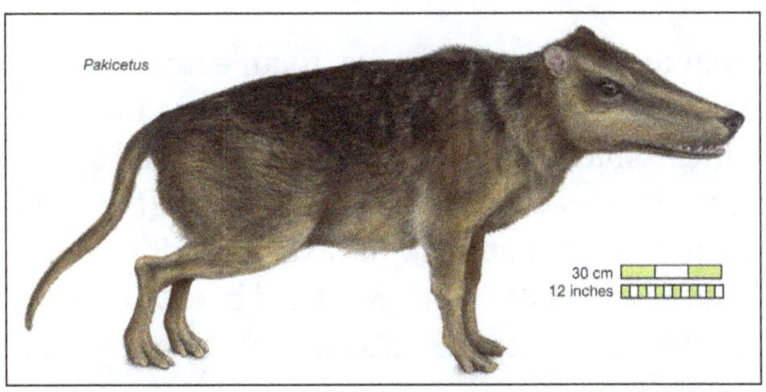

Pakicetus, an ancestral whale, the first cetacean discovered with functional legs

As the fossil record becomes more complete, the pattern will emerge as to which condylarth is ancestral to archaeocetes and which archaeocete is ancestral to living cetaceans. The first fossil cetacean, Pakicetus, is known from the Early Eocene Epoch (55.8 million to 49 million years ago) in Pakistan. It has recently become clear that archaeocetes rapidly diversified during the Eocene, and at least five now-extinct families are recognized. One subfamily of the Basilosauridae, the Dorudontinae, is thought to have given rise to both living suborders of cetaceans (baleen whales and toothed whales) sometime during the Late Oligocene Epoch. Evidence of the evolution of bony structures involved in echolocation appears in toothed whales from the Late Oligocene. The earliest toothed whale to display them, Echovenator sandersi, emerged some 27 million years ago.

The first baleen whales had wide, flat skulls bordered by a reduced number of teeth in the archaeocete pattern. The roof of the mouth widened between these borders, and grooves for blood vessels supplying the emerging baleen are seen inside the tooth rows. By the middle of the Miocene Epoch (some 16 million to 11 million years ago), there were several families of baleen whales, including the right whales and rorquals. The Miocene was the epoch during which modern ocean circulation began; regional areas of upwelling and increased productivity developed, setting the stage for the evolution of large whales with seasonal migratory distribution. At the same time, the modern toothed whales began to emerge, developing into nine families during this period; four of these have since become extinct. Sperm whales were among the first toothed whales and were present during the Middle Miocene as large and well-defined as they are now. From the fossil record it is evident that today's cetacean biodiversity has decreased markedly since the Miocene.

Classification

Though there is some consensus among taxonomists that Cetacea should be treated as one order, others believe they are actually two or three. This depends on the evaluation of the degree of shared ancestry, which remains controversial. It is possible that certain protocetids, or "pre-whales," could have given rise to both modern groups of whales, which is why some authorities prefer a single order. However, the absence of intermediate fossils linking baleen whales with toothed forms supports the use of separate orders. Resolution of this problem awaits the discovery of relevant fossils.

Further disagreement occurs at many points below the ordinal level. Although there is no doubt that any recent cetacean is either toothed (suborder Odontoceti) or baleen (suborder Mysticeti), the relationships of many genera are in doubt. For example, the long-snouted dolphins are classified by some authorities as a separate family, Stenidae, rather than family Delphinidae. A similar situation exists for the porpoises (family Phocoenidae). At the species level there is uncertainty about the specific or subspecific status of many populations.

It must be borne in mind that all classifications are, to an extent, artificial. Over time one species merges with another and some classification issues actually get more complex as the fossil record improves. Current cetacean classification is the result of an improving fossil record that reveals more taxa near the origins of the three established suborders (the archaeocetes, mysticetes, and odontocetes). It has become apparent that a major diversification was associated with each major adaptive branch, and species are currently being shuttled back and forth between the latest representatives of one family and the earliest representatives of its descendant family.

Although species relationships have historically been based on morphological (anatomic) characteristics, biologists have begun comparing DNA sequences of different cetaceans and thereby causing morphologists to reexamine their taxonomic data. Chromosome count (karyotype) can be quite variable between related mammals, but it is remarkably stable among cetaceans. All baleen whales and most toothed whales have 44 chromosomes, and sperm whales and beaked whales have 42. Techniques such as DNA sequencing have provided different ways of evaluating the relationships of species. Applications of these and older techniques should provide a clearer understanding of the evolutionary history of Cetacea.

Annotated Taxonomy

The three cetacean suborders (two living, one fossil) recognized below share the same basic body plan but differ in their degree of specialization. Suborders and families are separated primarily on the basis of the following characteristics: tooth structure, number, and degree of differentiation;

skull modifications, especially the position of the nostrils, degree of telescoping of the whole skull, modifications to the inner and middle ear, and extent of joining of the two halves of the jaw; and degree of modification of the pelvic girdle (archaeocetes only).

Pinnipeds

Pinnipeds are are large aquatic (mostly marine) mammals having both front and hind limbs in the form of flippers and comprising the taxonomic group Pinnipedia. Pinnipedia is usually considered a suborder of the order Carnivora, but is sometimes considered a separate order; more recently, Pinnipedia is regarded as a superfamily of the suborder Caniformia of Carnivora.

Pinnipeds include walruses, earless seals (or true seals), and eared seals (sea lions and fur seals). They are one of four groups of marine mammals, the others being cetaceans (whales, dolphins, and porpoises), sirenians (manatees and dugongs), and sea otters. The cetaceans and sirenians are completely aquatic, however, and the sea otters can mate and raise their young entirely at sea, while the pinnipeds spend considerable time on land, including giving birth and raising their young.

Pinnipeds, reflecting the wide diversity in nature, exhibit diverse reproductive behaviors and occupy a wide variety of habitats, from the tropics to the polar region. Seals are almost exclusively marine, but one species, the Baikal seal or nerpa (Phoca sibirica) lives in Lake Baikal, a freshwater environment, and two subspecies of the ringed seal also spend their whole lives in freshwater.

Taxonomy

The placement of Pinnipedia in taxonomic schemes has been a source of controversy. Some systematists regard them as a separate order based on their uniqueness as a group. Generally, however, they are placed within the order Carnivora.

Carnivora is an order of placental animals that includes such familiar groups as bears, foxes, cats, hyenas, skunks, otters, raccoons, and so forth. Older classification schemes divided this order into Pinnipedia (marine carnivores) and Fissipedia (primarily land carnivores). However, pinnipeds are considered to have derived from land-based Carnivora ancestors, making this an unnatural division from the point of view of descent.

Thus, the division into Pinnipedia and Fissipedia has given way to taxonomic schemes whereby pinnipeds are placed as a subdivision of Carnivora associated with or within the arctoid group ("bear-like" carnivores). Newer classification schemes divide the Carnivora into the suborders Caniformia (dog-like) and Feliformia (cat-like). Pinnipeds are placed in the Caniformia group, generally as a superfamily.

There is greater consensus on the division of pinnipeds into three families. These are:

- Odobenidae – walruses
- Otariidae – sea lions, eared seals, and fur seals
- Phocidae – true (earless) seals

Odobenidae and Otariidae are generally placed together in the superfamily Otarioidea, while Phocidae is in the superfamily Phocoidea. However, some consider walruses more closely related to the true seals. McKenna and Bell place walruses with the true seals in the family Phocidae, a practice that is followed by the Tree of Life Web Project, which yields the following two families:

- Otariidae – sea lions, eared seals, and fur seals.
- Phocidae – true (earless) seals and walruses.

One very visible difference between the otariids and the phocids is the pinna, a small furry earflap, found on the otariids. Phocids are thus referred to as "earless seals," since their ears are not easily seen, while otariids are referred to as "eared seals." Another difference is that otariids have hindflippers that can be inverted under the body, aiding their movement on land, while the hindflippers of phocids cannot be turned forward under the body, causing their movement on land to be slow and awkward. Otariids also swim using their long front flippers to move themselves through the water, while phocids swim by using their rearflippers and lower body in a side-to-side motion. The breeding system of the two groups is also different.

There are 33 or 34 recognized species of pinnipeds. These include one species of odobenid (walrus), 14 otariids, and 18 or 19 phocids (one, the Caribbean monk seal, may be extinct).

Distribution

There are an estimated 50 million pinnipeds in the world; roughly 90 percent of them are phocid seals. Among the pinnipeds, only phocids (true seals) live in the Antarctic and Arctic, while walruses inhabit the northern circumpolar waters; there are no otariids in the extreme polar regions.

Only two species of phocids are found in the tropics and these are small and endangered populations, but a number of fur seals and sea lions live in tropical and subtropical areas, with fur seals ranging widely into colder climates as well. All but two species of fur seals are found in the Southern Hemisphere, while sea lions are common to both hemispheres.

No Pinnipeds are Found in Asia

Although fewer species of pinnipeds inhabit the Antarctic than the Arctic, the Antarctic seal populations are larger (including about 30 million or so crab-eater seals); in addition, the Antarctic seals also generally have larger bodies. The larger size may be an adaptation providing more insulation or may be related to more plentiful food supply.

Description

Pinnipeds are typically sleek-bodied and rather large. Their bodies are well adapted to their aquatic habitat, in which they spend most of their lives. In place of hands, their forelimbs are large flippers (hence the name "fin-feet"), and their bodies narrow out into a tail.

The smallest pinniped, the Galapagos fur seal, weighs about 30 kg (66 lb) when full-grown and is 1.2 m (4 ft) long; the largest, the male southern elephant seal, is over 4 m (13 ft) long and weighs up to 2,200 kg (4,850 lb, more than 2 tons).

Pinnipeds are more streamlined, larger, and faster swimming than the sea otters, which have retained the front paws of a terrestrial carnivore, although their hind feet are flippers.

All pinnipeds are carnivorous, eating fish, shellfish, squid, and other marine creatures. The leopard seal is probably the most carnivorous and predatory of all the pinnipeds, eating a wide variety of prey, from krill to penguins to other seals.

Reproduction

Unlike sea otters, which, although they can crawl on land, essentially retain no ties to land and are able to mate and raise their offspring entirely at sea, the pinnipeds are still dependent on land and return to land to give birth and nurse their young.

Otariid seals tend to have similar breeding systems, with all sea lions and fur seals being highly polygynous. Adult males are normally territorial and go without eating during the breeding season. Sexual dimorphism is pronounced among otariid seals, with males two to four times the size of females. Phocid breeding behavior is more variable, with a variety of systems and variable degrees of sexual dimorphism. Some phocids breed in solitary groups, with just the male, female, and pup, and with the male and female of equal size; others form dominance hierarchies, with larger males competing for females; and there are species where the breeding takes place in large groups with hundreds of animals.

In general, male otariids defend a territory, whereas phocids with a dominance hierarchy defend clusters of females. These strategies reflect the different levels of mobility on land between otariids and phocids. As noted, the former have opposable hind flippers that can be placed flat on the ground to aid locomotion. True seals, on the other hand, cannot do this and drag themselves along using only their front flippers.

Pinnipeds come ashore to breed (haul-out), and this often necessitates traveling long distances from their feeding grounds to suitable mating grounds (either on land or ice). Because of these constraints, pinnipeds tend to mate and give birth with a high level of reproductive synchrony. The most synchronous species are the two phocids (seals)—the harp and hooded seals—in which all females are estimated to become sexually receptive during a period of 10–15 days. Males in dominance hierarchies compete for females at rookeries. Females are usually highly clustered here to reduce the level of male harassment, particularly by low-ranking males. Females compete for central positions and call out if attacked by subordinate males that are subsequently chased away by the dominant bull.

After giving birth, pinniped mothers suckle their young for a variable length of time. Unlike otariids, which nurse the pups for several months while alternating feeding at sea, most phocid females nurse for short periods while fasting or feeding little. Among the phocids, lactation varies from 4 to 50 days, whereas the otariids may lactate from 4 to 36 months. This reflects the fact that phocid feeding grounds tend to be a long way off-shore so lactation is associated with maternal fasting. To compensate for the short lactation period, the fat content of phocid milk is higher than in any other species of marine mammal (45–60 percent fat). After lactation, most female phocids make extensive migratory movements to feeding grounds for intensive foraging to recoup depleted energy reserves. On the other hand, otariid feeding grounds are generally closer to shore and females go on foraging trips to maintain lactation. Fat content of otariid milk is lower than that of the phocids owing to the protracted lactatory period (typically 25–50 percent). Protracted nursing also leads to the formation of social bonds.

Females have a postpartum oestrus allowing them to mate soon after giving birth. Subsequent implantation of the embryo is delayed (embryonic diapause) thus removing the need to come ashore (haul-out) twice, once to give birth and again later to mate. All otariids copulate soon after birth, while most phocids mate during late lactation or after weaning their pup.

Mustelidae

Mustelidae is one of the most successful families of the 5 order Carnivora; its 23 genera and 65 extant species of weasels, badgers, otters, and skunks have a nearly worldwide distribution. These small to medium-sized mammals tend to have long and slender bodies, exhibit sexual dimorphism (males are generally 25% larger than females), and possess well-developed anal scent glands. Several cranial and dental characters, including flooring of the suprameatal fossa (an expansion into the roof of the auditory canal), reduction/loss of the upper second molar (M2), reduction/loss of the meta-conule and lingual expansion of the upper first molar (Ml), and loss of the carnassial notch on the last upper premolar (P4), have been argued to be synapomorphies of a monophyletic Mustelidae. Recent molecular analyses, however, suggest that skunks are distantly related to other mustelids and occupy a phylogenetic position as sister taxon to Musteloidea, a clade consisting of Mustelidae (minus skunks) and Procyonidae (raccoons and their relatives).

Palaeomusteloids, Early Eumustelids and Origins of Lutrinae

The earliest mustelid-like fossils are assigned to Mustelavus from the late Eocene of western North America and Mustelic-tis and Plesictis from the early Oligocene of France. Because these animals retain upper and lower second molars and an upper carnassial notch, they are conservatively placed in an undifferentiated musteloid stem group. Others have argued, however, that the small size of the second molars and reduction of the metaconule are indicative of mustelid affinities and place all three genera in the family Mustelidae. An increased diversity of palaeomusteloids occurred in the late Oligocene-early Miocene of North America and Europe, and although most of these taxa have little to do with later mustelid evolution, two European genera, Paragale and Plesiogale, lose the carnassial notch, exhibit some flooring of the suprameatal fossa, and show rudimentary lingual expansion of the upper first molar suggesting an ancestral relationship to eumustelids (Hunt, 1996). Although eumustelids evolved in Eurasia, the earliest representatives of this group are found in the early Miocene of North America, suggesting a late Oligocene immigration event, with immigration events into Africa and South America occurring in the early Miocene and late Pliocene, respectively. By the late Miocene there is considerable mustelid diversity in the fossil record and all of the extant mustelid subfamilies are represented.

The dentitions of a canid (gray fox, A) and mustelid (fisher, B) are compared in lateral (top) and occlusal views (bottom). Eumustelids have lost the second molar, the metaconule on Ml and the carnassial notch on P4, and expanded the lingual surface of Ml. Scale bar = 2 cm.

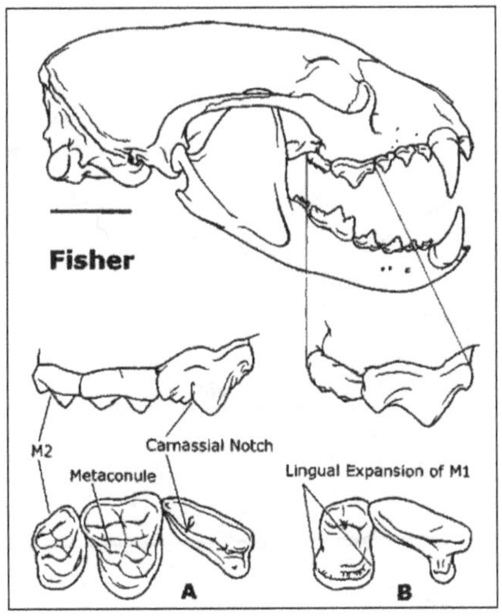

Characteristics of upper carnassial (P4) and molar (Ml) teeth in mustelids

One enigmatic taxon that has been included with the palaeomusteloids by some workers and as an incertae sedis member of Arctoidea [Musteloidea + Ursidae (bears) + Pinnipedia] by others is the late Oligocene Potamotherium, Known from a number of complete and well-preserved skeletons, this aquati-cally adapted otter-like genus exhibits a combination of musteloid and pinniped characters that have suggested to some that it may be transitional between mustelids and phocids. although others have argued that these morphological similarities are convergent. The earliest widely accepted fossil otter is the early to mid-Miocene genus Mio-nictis known from dental remains in North America, Europe, and China. Diversification of the major lutrine clades likely occurred by the mid-Miocene, although fossil records for most lineages tend to be Plio-Pleistocene in age. One exception to this is the fossil record of the sea otter (Enhydra lutris), the most marine and one of the most morphologically derived of all otters. Remains of Enhy-dritherium and Enhydriodon, the consecutive outgroups to Enhydra, respectively, have been described from throughout the late Miocene, sharing with the living sea otter loss of upper and lower first premolars, a short robust jaw, and lower first molars with low, inflated cusps.

Phylogenetic Relationships of Musteloids

A number of recent morphologic and molecular analyses have addressed phylogenetic relationships both among genera that have been referred to Mustelidae and between mustelids and other extant members of the monophyletic suborder Can-iformia, which includes canids (dogs), procyonids. pinnipeds, and ursids. While most of these analyses have supported a traditionally held view that Procyonidae and Mustelidae are sister taxa, and that Ursidae and Canidae represent successive out-groups, the position of Pinnipedia has proven more difficult to resolve, with morphological evidence supporting a sister taxon relationship to Ursidae and molecular data tending to support a sister group relationship to Musteloidea.

Cladograms depicting the relationship of mustelids to other extant members of the carnivoran suborder Caniformia (A) and one hypothesis of phylogenetic relationships among musteloids based on molecular data where skunks (Mephitinae) are not part of a monophyletic Mustelidae. Dotted lines (A) show alternative hypotheses for pinniped relationships, and the musteline/meline designation (B) is used to indicate the para-phyletic nature of these subfamilies.

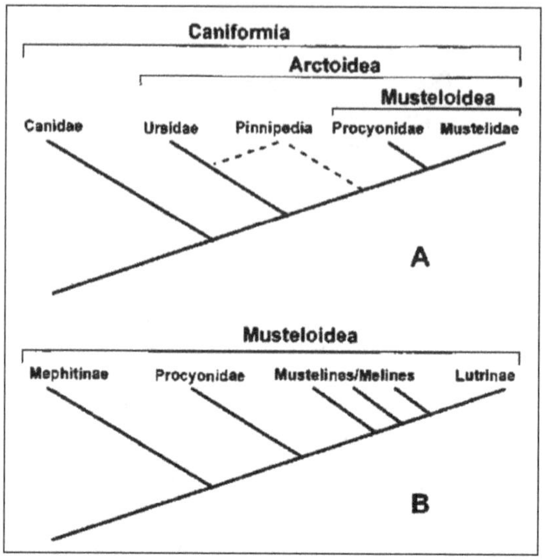

Among the 23 genera and five subfamilies attributed to Mustelidae, there is strong support, both molecular and morphologic, for the monophyly of otters (Lutrinae) and skunks (Mephitinae), little support for Mustelinae monophyly, and no support for badger (Melinae) monophyly (weasels, martens, wolverines, among others). Lutrinae appears to be nested well within the mustelid clade. and several studies have suggested that skunks are the sister taxon to otters. Ribosomal protein sequence data, however, argue for a phylogenetic position of skunks as sister taxon to Musteloidea. implying that characters such as loss of the second molars and upper carnas-sial notch evolved more than once among musteloids. Consensus on musteloid phylogeny will require more analyses that combine molecular and morphologic data, and that include important fossil taxa such as Mustelavus and Potamotherium. as well as the early procyonid Amphictis.

Sirenians

Sirenia is an order of aquatic mammals, represented today by three species of manatees (family Trichechidae) and one species of dugong (family Dugongidae). Commonly referred to as "sea cows" for their grazing of vegetation, members of this order also are known collectively as sirenians and sirens. All four extant species are large, herbivorous, and fully aquatic, inhabiting swamps, rivers, estuaries, marine wetlands, and coastal marine waters. These mammals all have nearly hairless skin, paddle-like front flippers, ears without pinnae, and effectively no hind limbs, being reduced to only a vestigial pelvis.

Manatees and dugongs are the only marine mammals classified as herbivores. The dugong is the only strictly-marine herbivorous sirenian, as all species of manatee utilize freshwater to some degree. The dugong is found in Indo-Pacific waters from eastern Africa to southeast Asia, while the manatees are found in both freshwater and marine environments in parts of the Americas and Africa. One representative, the Steller's sea cow (Hydrodamalis gigas) became extinct only since the 18th century.

As the only marine mammals herbivores, sirenians fill a unique niche in food chains, being the largest primary consumer of aquatic vegetation, such as seagrass. However, the also have been hunted for food, leather, and other products, and their habitat developed for residential, commercial, and recreation use, to the point where populations have been severely limited. The three manatee species (family Trichechidae) and the dugong (family Dugongidae) are endangered species. All four are vulnerable to extinction from habitat loss and other negative impacts related to human population growth and coastal development, including destruction of seagrass beds by increased sedimentation. Steller's sea cow (Hydrodamalis gigas) provides a particularly sobering reflection on the severity of human impacts. Steller's sea cow has been extinct since 1786, just 27 years after its discovery in 1741, due largely to hunting by sailors/explorers provisioning their ships for long voyages.

Sirenians are one of four extant groups of marine mammals, the others being cetaceans (whales, dolphins, and porpoises), sea otters, and pinnipeds (walruses, earless seals, and eared seals). The sirenians and cetaceans are completely aquatic, while pinnipeds spend considerable time on land, including giving birth and raising their young, and sea otters can mate and raise their young entirely at sea. Like dolphins and whales, manatees and dugongs are completely aquatic mammals that never leave the water—not even to give birth. Sirenia is thought to have evolved from four-legged land mammals over 60 million years ago, with the closest living relatives being the Proboscidea (elephants) and Hyracoidea (hyraxes).

There are two extant families and four species in Sirenia: Trichechidae (three species of manatees, Trichechus inunguis, T. manatus, and T. senegalensis) and Dugongidae (one species, the dugong, Dugong dugon). Sirenia also include Steller's sea cow, extinct since the 18th century, and a number of taxa known only from fossils. The order evolved during the Eocene, more than 50 million years ago. About 50 species in 4 families have been described in total as of 2006. The oldest fossil records of sirenians, from the Middle Eocene, were placed in the family Prorastomidae and were amphibious quadrupeds.

Sirenia, commonly "sirenians", are also referred to by the common name "sirens", deriving from the sirens of Greek mythology, which also were confused by authors and sailors with the mythical mermaids.

Modern sirenians have major aquatic adaptations: the forelimbs have modified into paddle-like front flippers used for steering, the tail has modified into a paddle used for propulsion, and the hindlimbs (legs) are lacking with the exception of two small remnant (vestigial) bones floating deep in the muscle. Although sirenians appear fat, they are fusiform, hydrodynamic, and highly muscular. Their skulls are highly modified for taking breaths of air at the water's surface, and dentition is greatly reduced. Although mammals, they have nearly hairless bodies, and they lack external ears and have relatively small eyes. The nostrils, located on the top of their snouts, are closed by values, while their eyes are closed by a sphincter-like mechanism. Their large and mobile lips are covered by stiff bristles.

Sirenians also experience pachyostosis, a condition in which the ribs and other long bones are unusually solid and contain little or no marrow. These heavy bones, which are among the densest in the animal kingdom, may act as a ballast, helping to neutralize the buoyancy of their blubber and keeping sirenians suspended slightly below the water's surface.

The lungs of sirenians are unlobed. In sirenians, the lungs and diaphragm extend the entire length of the vertebral column. These adaptations help sirenians control their buoyancy and maintain their horizontal position in the water.

The premaxillae of the skull are large and downward deflected and the tympanic bone in the region of the ear is semicircular. Sirenians have only two teats, located under their forelimbs, similar to elephants. The elephants are thought to be the closest living relatives of the sirenians.

The manatee appears to have an almost unlimited ability to produce new teeth as the anterior teeth wear down. Unlike the manatees, the dugong's teeth do not continually grow back via horizontal tooth replacement.

The main visual difference between manatees and dugongs are the tails. A manatee tail is paddle-shaped, while a dugong tail is forked, similar in shape to that of a whale of dolphin. The fluked tail provides locomotion through vertical movement. The Trichechidae species differ from the Dugongidae also in the shape of their skull.

Living sirenians grow between 2.5 and 4.0 meters long and can weigh up to 1,500 kg. Hydrodamalis gigas, Steller's sea cow, was huge (size of African elephants in weight, over 6000 kg) and could reach lengths of 8 m. Dugongs are generally smaller than manatees (with the exception of the Amazonian manatee), reaching an average adult length of 2.4 to 3 meters (8 to 10 feet) and a weight of 231 to 499 kilograms (510 to 1100 pounds). An adult's length rarely exceeds 3 meters, and females tend to be larger than males. The largest known dugong was an exceptional female landed off the Saurashtra coast of west India, measuring 4.03 meters (13.3 feet) and weighing 1,018 kilograms (2,240 pounds).

Feeding

Manatees and dugongs are the only marine mammals classified as herbivores. Unlike the other marine mammals (dolphins, whales, seals, sea lions, sea otters, and walruses), sirenians eat primarily sea grasses and other aquatic vegetation, and have an extremely low metabolism and poor tolerance for especially cold water. Sirenians have been observed eating dead animals (sea gulls), but their diets are made up primarily of vegetation. These animals have been observed eating grass clippings from homes adjacent to waterways, but in this rare occurrence, only the top portion of the sirenian is lifted out of the water. The combination of these factors means sirenians are restricted to warm, shallow, coastal waters, estuaries, and rivers with healthy ecosystems that support large amounts of seagrass or other vegetation.

The dugong is heavily dependent on seagrasses for subsistence. A "cultivation grazer," it feeds on the seagrass in a manner that actually promotes regrowth of the seagrass, though aerating the sea floor and increasing the amount of organic matter in the area. The huge Steller's sea cow, which lived in the Bering sea, feed exclusively on seaweed.

Polar Bears

Polar bears are classified as marine mammals because they spend most of their lives on the sea ice of the Arctic Ocean. They have a thick layer of body fat and a water-repellant coat that insulates them from the cold air and water. Considered talented swimmers, they can sustain a pace of six miles per hour by paddling with their front paws and holding their hind legs flat like a rudder.

Polar bears spend over 50% of their time hunting for food. A polar bear might catch only one or two out of ten seals it hunts, depending on the time of year and other variables. Their diet mainly consists of ringed and bearded seals because they need large amounts of fat to survive.

Polar bears are aggressive predators, known for taking a variety of prey. They spend most of the year associated with Arctic sea ice, where they hunt ringed seal and bearded seal pups. Polar bears have an excellent sense of smell and can locate pups even when they are buried in snow dens. Polar bears can easily use their size and strength to break open dens and feed on the helpless pups. Adult males have also been known to eat large aquatic animals, like beluga whales, that they attack from the ice and pull onto the ice surface. When on land (as opposed to surface ice), polar bears scavenge for dead animal matter. This species has no natural predator. It is one of the few species that is even considered to be higher on the food web than humans.

Courtship and mating take place on the ice surface, but birth generally takes place on land. Females give birth to (typically) two cubs, which weigh no more than a few pounds (~one kilogram). Cubs nurse for at least two and a half years, and as in all bears, the mother polar bear is aggressively protective of her offspring. Polar bears reach sexual maturity between ages four and six.

Polar bear populations are decreasing throughout their geographic range, and some populations have been depleted significantly. Scientists now believe the species to be vulnerable to extinction. While hunting for this species does occur, climate change is the primary threat to the polar bear's existence. This bear is intimately tied to Arctic sea ice and to seals that require ice for reproduction, and that ice is expected to continue thinning and melting faster each summer in the face of climate change. Scientists are not sure if polar bears will be able to adapt to these rapidly changing conditions.

References

- Marine-mammals, marine-life: noaa.gov, Retrieved 10 March, 2019
- Marine-mammals, marine-vertebrates: marinebio.org, Retrieved 8 February, 2019
- Adaptations: uaf.edu, Retrieved 23 July, 2019
- Cetacean, animal: britannica.com, Retrieved 3 April, 2019
- Cetacean: newworldencyclopedia.org, Retrieved 11 February, 2019
- Natural-history, cetacean, animal: britannica.com, Retrieved 17 April, 2019
- Pinniped: newworldencyclopedia.org, Retrieved 23 May, 2019
- Mustelidae-marine-mammals, marine-mammals: what-when-how.com, Retrieved 19 August, 2019
- Polar-bear, species: worldwildlife.org, Retrieved 14 June, 2019
- Polar-bear, marine-mammals, marine-life: oceana.org, Retrieved 20 January, 2019

5

Marine Ecosystem

The primary characteristic of a marine ecosystem is water that has a high salt content. There are a number of marine ecosystems such as coral reef ecosystem, mangrove ecosystem, kelp forest ecosystems and rocky shore ecosystem. The chapter closely examines these key types of marine ecosystems to provide an extensive understanding of the subject.

Marine ecosystem is complex of living organisms in the ocean environment. Marine waters cover two-thirds of the surface of the Earth. In some places the ocean is deeper than Mount Everest is high; for example, the Mariana Trench and the Tonga Trench in the western part of the Pacific Ocean reach depths in excess of 10,000 metres (32,800 feet). Within this ocean habitat live a wide variety of organisms that have evolved in response to various features of their environs.

The Earth formed approximately 4.5 billion years ago. As it cooled, water in the atmosphere condensed and the Earth was pummeled with torrential rains, which filled its great basins, forming seas. The primeval atmosphere and waters harboured the inorganic components hydrogen, methane, ammonia, and water. These substances are thought to have combined to form the first organic compounds when sparked by electrical discharges of lightning. Some of the earliest known organisms are cyanobacteria. Evidence of these early photosynthetic prokaryotes has been found in Australia in Precambrian marine sediments called stromatolites that are approximately 3 billion years old. Although the diversity of life-forms observed in modern oceans did not appear until much later, during the Precambrian (about 4.6 billion to 542 million years ago) many kinds of bacteria, algae, protozoa, and primitive metazoa evolved to exploit the early marine habitats of the world. During the Cambrian Period (about 542 million to 488 million years ago) a major radiation of life occurred in the oceans. Fossils of familiar organisms such as cnidaria (e.g., jellyfish), echinoderms (e.g., feather stars), precursors of the fishes (e.g., the protochordate Pikaia from the Burgess Shale of Canada), and other vertebrates are found in marine sediments of this age. The first fossil fishes are found in sediments from the Ordovician Period (about 488 million to 444 million years ago). Changes in the physical conditions of the ocean that are thought to have occurred in the Precambrian—an increase in the concentration of oxygen in seawater and a buildup of the ozone layer that reduced dangerous ultraviolet radiation—may have facilitated the increase and dispersal of living things.

Marine Environment

Geography, Oceanography and Topography

The shape of the oceans and seas of the world has changed significantly throughout the past 600 million years. According to the theory of plate tectonics, the crust of the Earth is made up of many dynamic plates. There are two types of plates—oceanic and continental—which float on the surface of the Earth's mantle, diverging, converging, or sliding against one another. When two plates diverge, magma from the mantle wells up and cools, forming new crust; when convergence occurs, one plate descends—i.e., is subducted—below the other and crust is resorbed into the mantle. Examples of both processes are observed in the marine environment. Oceanic crust is created along oceanic ridges or rift areas, which are vast undersea mountain ranges such as the Mid-Atlantic Ridge. Excess crust is reabsorbed along subduction zones, which usually are marked by deep-sea trenches such as the Kuril Trench off the coast of Japan.

The shape of the ocean also is altered as sea levels change. During ice ages a higher proportion of the waters of the Earth is bound in the polar ice caps, resulting in a relatively low sea level. When the polar ice caps melt during interglacial periods, the sea level rises. These changes in sea level cause great changes in the distribution of marine environments such as coral reefs. For example, during the last Pleistocene Ice Age the Great Barrier Reef did not exist as it does today; the continental shelf on which the reef now is found was above the high-tide mark.

Marine organisms are not distributed evenly throughout the oceans. Variations in characteristics of the marine environment create different habitats and influence what types of organisms will inhabit them. The availability of light, water depth, proximity to land, and topographic complexity all affect marine habitats.

The availability of light affects which organisms can inhabit a certain area of a marine ecosystem. The greater the depth of the water, the less light can penetrate until below a certain depth there is no light whatsoever. This area of inky darkness, which occupies the great bulk of the ocean, is called the aphotic zone. The illuminated region above it is called the photic zone, within which are distinguished the euphotic and disphotic zones. The euphotic zone is the layer closer to the surface that receives enough light for photosynthesis to occur. Beneath lies the disphotic zone, which is illuminated but so poorly that rates of respiration exceed those of photosynthesis. The actual depth of these zones depends on local conditions of cloud cover, water turbidity, and ocean surface. In general, the euphotic zone can extend to depths of 80 to 100 metres and the disphotic zone to depths of 80 to 700 metres. Marine organisms are particularly abundant in the photic zone, especially the euphotic portion; however, many organisms inhabit the aphotic zone and migrate vertically to the photic zone every night. Other organisms, such as the tripod fish and some species of sea cucumbers and brittle stars, remain in darkness all their lives.

Marine environments can be characterized broadly as a water, or pelagic, environment and a bottom, or benthic, environment. Within the pelagic environment the waters are divided into the neritic province, which includes the water above the continental shelf, and the oceanic province, which includes all the open waters beyond the continental shelf. The high nutrient levels of the neritic province—resulting from dissolved materials in riverine runoff distinguish this province from the oceanic. The upper portion of both the neritic and oceanic waters the epipelagic zone is where photosynthesis occurs; it is roughly equivalent to the photic zone. Below this zone lie the mesopelagic, ranging between 200 and 1,000 metres, the bathypelagic, from 1,000 to 4,000 metres, and the abyssalpelagic, which encompasses the deepest parts of the oceans from 4,000 metres to the recesses of the deep-sea trenches.

The benthic environment also is divided into different zones. The supralittoral is above the high-tide mark and is usually not under water. The intertidal, or littoral, zone ranges from the high-tide mark (the maximum elevation of the tide) to the shallow, offshore waters. The sublittoral is the environment beyond the low-tide mark and is often used to refer to substrata of the continental shelf, which reaches depths of between 150 and 300 metres. Sediments of the continental shelf that influence marine organisms generally originate from the land, particularly in the form of riverine runoff, and include clay, silt, and sand. Beyond the continental shelf is the bathyal zone, which occurs at depths of 150 to 4,000 metres and includes the descending continental slope and rise. The abyssal zone (between 4,000 and 6,000 metres) represents a substantial portion of the oceans. The deepest region of the oceans (greater than 6,000 metres) is the hadal zone of the deep-sea trenches. Sediments of the deep sea primarily originate from a rain of dead marine organisms and their wastes.

Physical and Chemical Properties of Seawater

The physical and chemical properties of seawater vary according to latitude, depth, nearness to land, and input of fresh water. Approximately 3.5 percent of seawater is composed of dissolved compounds, while the other 96.5 percent is pure water. The chemical composition of seawater reflects such processes as erosion of rock and sediments, volcanic activity, gas exchange with the atmosphere, the metabolic and breakdown products of organisms, and rain. In addition to carbon, the nutrients essential for living organisms include nitrogen and phosphorus, which are minor constituents of seawater and thus are often limiting factors in organic cycles of the ocean. Concentrations of phosphorus and nitrogen are generally low in the photic zone because they are rapidly taken up by marine organisms. The highest concentrations of these nutrients generally are found below 500 metres, a result of the decay of organisms. Other important elements include silicon and calcium (essential in the skeletons of many organisms such as fish and corals).

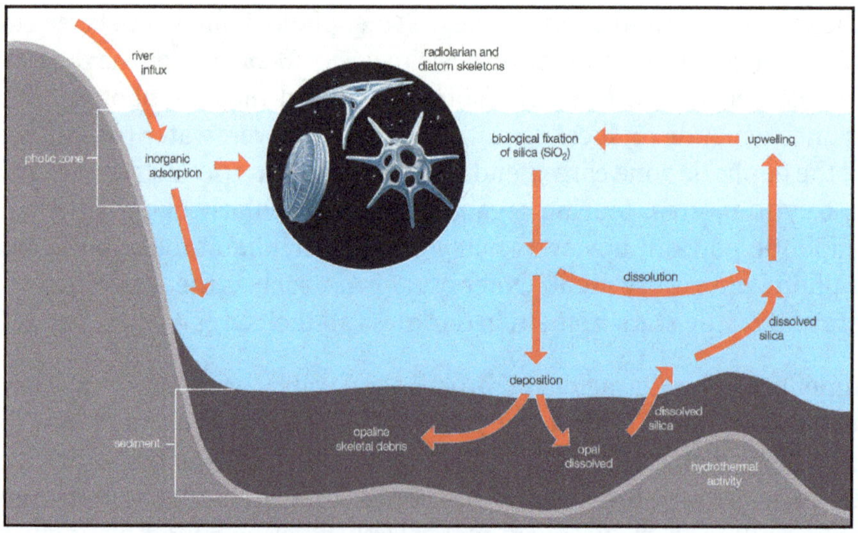

Cycling of silica in the marine environment - Silicon commonly occurs in nature as silicon dioxide (SiO_2), also called silica. It cycles through the marine environment, entering primarily through riverine runoff. Silica is removed from the ocean by organisms such as diatoms and radiolarians that use an amorphous form of silica in their cell walls. After they die, their skeletons settle through the water column and the silica redissolves. A small number reach the ocean floor, where they either remain, forming a silaceous ooze, or dissolve and are returned to the photic zone by upwelling.

The chemical composition of the atmosphere also affects that of the ocean. For example, carbon dioxide is absorbed by the ocean and oxygen is released to the atmosphere through the activities of marine plants. The dumping of pollutants into the sea also can affect the chemical makeup of the ocean, contrary to earlier assumptions that, for example, toxins could be safely disposed of there.

The physical and chemical properties of seawater have a great effect on organisms, varying especially with the size of the creature. As an example, seawater is viscous to very small animals (less than 1 millimetre - 0.039 inch long) such as ciliates but not to large marine creatures such as tuna.

Marine organisms have evolved a wide variety of unique physiological and morphological features that allow them to live in the sea. Notothenid fishes in Antarctica are able to inhabit waters as cold as −2 °C (28 °F) because of proteins in their blood that act as antifreeze. Many organisms are able to achieve neutral buoyancy by secreting gas into internal chambers, as cephalopods do, or into swim bladders, as some fish do; other organisms use lipids, which are less dense than water, to achieve this effect. Some animals, especially those in the aphotic zone, generate light to attract prey. Animals in the disphotic zone such as hatchetfish produce light by means of organs called photophores to break up the silhouette of their bodies and avoid visual detection by predators. Many marine animals can detect vibrations or sound in the water over great distances by means of specialized organs. Certain fishes have lateral line systems, which they use to detect prey, and whales have a sound-producing organ called a melon with which they communicate. Tolerance to differences in salinity varies greatly: stenohaline organisms have a low tolerance to salinity changes, whereas euryhaline organisms, which are found in areas where river and sea meet (estuaries), are very tolerant of large changes in salinity. Euryhaline organisms are also very tolerant of changes in temperature. Animals that migrate between fresh water and salt water, such as salmon or eels, are capable of controlling their osmotic environment by active pumping or the retention of salts. Body architecture varies greatly in marine waters. The body shape of the cnidarian by-the-wind-sailor (velella)—an animal that lives on the surface of the water (pleuston) and sails with the assistance of a modified flotation chamber—contrasts sharply with the sleek, elongated shape of the barracuda.

Ocean Currents

The movements of ocean waters are influenced by numerous factors, including the rotation of the Earth (which is responsible for the Coriolis Effect), atmospheric circulation patterns that influence surface waters, and temperature and salinity gradients between the tropics and the polar regions (thermohaline circulation). The resultant patterns of circulation range from those that cover great areas, such as the North Subtropical Gyre, which follows a path thousands of kilometres long, to small-scale turbulences of less than one metre.

Marine organisms of all sizes are influenced by these patterns, which can determine the range of a species. For example, krill (Euphausia superb) are restricted to the Antarctic Circumpolar Current. Distribution patterns of both large and small pelagic organisms are affected as well. Mainstream currents such as the Gulf Stream and East Australian Current transport larvae great distances. As a result cold temperate coral reefs receive a tropical infusion when fish and invertebrate larvae from the tropics are relocated to high latitudes by these currents. The successful recruitment of eels to Europe depends on the strength of the Gulf Stream to transport them from spawning sites in the Caribbean. Areas where the ocean is affected by nearshore features, such as estuaries, or areas in which there is a vertical salinity gradient (halocline) often exhibit intense biological activity. In these environments, small organisms can become concentrated, providing a rich supply of food for other animals.

Marine Biota

Marine biota can be classified broadly into those organisms living in either the pelagic environment (plankton and nekton) or the benthic environment (benthos). Some organisms, however, are benthic in one stage of life and pelagic in another. Producers that synthesize organic molecules exist in both environments. Single-celled or multicelled plankton with photosynthetic pigments are the producers of the photic zone in the pelagic environment. Typical benthic producers are microalgae (e.g., diatoms), macroalgae (e.g., the kelp Macrocystis pyrifera), or sea grass (e.g., Zostera).

Plankton

Plankton are the numerous, primarily microscopic inhabitants of the pelagic environment. They are critical components of food chains in all marine environments because they provide nutrition for the nekton (e.g., crustaceans, fish, and squid) and benthos (e.g., sea squirts and sponges). They also exert a global effect on the biosphere because the balance of components of the Earth's atmosphere depends to a great extent on the photosynthetic activities of some plankton.

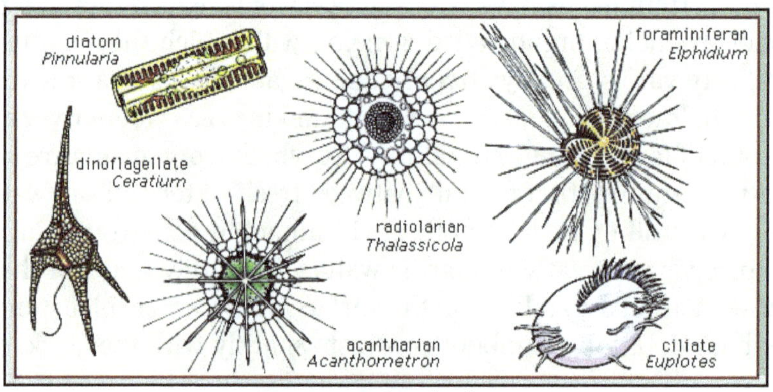

Representative plankton

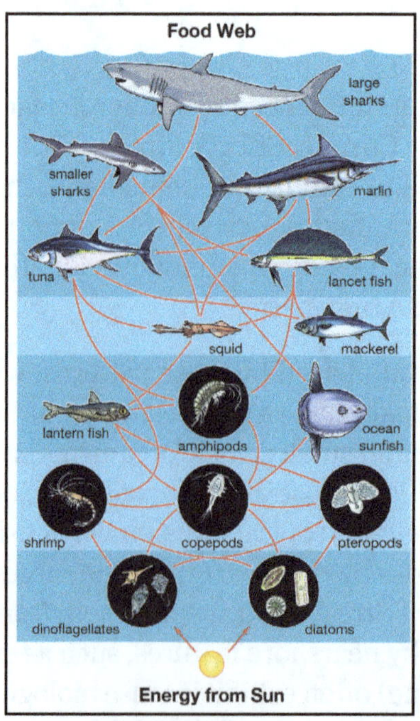

Generalized aquatic food web

The term plankton is derived from the Greek planktos, meaning wandering or drifting, an apt description of the way most plankton spend their existence, floating with the ocean's currents. Not all plankton, however, are unable to control their movements, and many forms depend on self-directed motions for their survival.

Plankton range in size from tiny microbes (1 micrometre [0.000039 inch] or less) to jellyfish whose gelatinous bell can reach up to 2 metres in width and whose tentacles can extend over 15 metres. However, most planktonic organisms, called plankters, are less than 1 millimetre (0.039 inch) long. These microbes thrive on nutrients in seawater and are often photosynthetic. The plankton include a wide variety of organisms such as algae, bacteria, protozoans, the larvae of some animals, and crustaceans. A large proportion of the plankton are protists—i.e., eukaryotic, predominantly single-celled organisms. Plankton can be broadly divided into phytoplankton, which are plants or plantlike protists; zooplankton, which are animals or animal-like protists; and microbes such as bacteria. Phytoplankton carry out photosynthesis and are the producers of the marine community; zooplankton are the heterotrophic consumers.

Diatoms and dinoflagellates (approximate range between 15 and 1,000 micrometres in length) are two highly diverse groups of photosynthetic protists that are important components of the plankton. Diatoms are the most abundant phytoplankton. While many dinoflagellates carry out photosynthesis, some also consume bacteria or algae. Other important groups of protists include flagellates, foraminiferans, radiolarians, acantharians, and ciliates. Many of these protists are important consumers and a food source for zooplankton.

Zooplankton, which are greater than 0.05 millimetre in size, are divided into two general categories: meroplankton, which spend only a part of their life cycle—usually the larval or juvenile stage—as plankton, and holoplankton, which exist as plankton all their lives. Many larval meroplankton in coastal, oceanic, and even freshwater environments (including sea urchins, intertidal snails, and crabs, lobsters, and fish) bear little or no resemblance to their adult forms. These larvae may exhibit features unique to the larval stage, such as the spectacular spiny armour on the larvae of certain crustaceans (e.g., Squilla), probably used to ward off predators.

Important holoplanktonic animals include such lobsterlike crustaceans as the copepods, cladocerans, and euphausids (krill), which are important components of the marine environment because they serve as food sources for fish and marine mammals. Gelatinous forms such as larvaceans, salps, and siphonophores graze on phytoplankton or other zooplankton. Some omnivorous zooplankton such as euphausids and some copepods consume both phytoplankton and zooplankton; their feeding behaviour changes according to the availability and type of prey. The grazing and predatory activity of some zooplankton can be so intense that measurable reductions in phytoplankton or zooplankton abundance (or biomass) occur. For example, when jellyfish occur in high concentration in enclosed seas, they may consume such large numbers of fish larvae as to greatly reduce fish populations.

The jellylike plankton are numerous and predatory. They secure their prey with stinging cells (nematocysts) or sticky cells (colloblasts of comb jellies). Large numbers of the Portuguese man-of-war (Physalia), with its conspicuous gas bladder, the by-the-wind-sailor (Velella *velella*), and the small blue disk-shaped Porpita porpita are propelled along the surface by the wind, and after strong onshore winds they may be found strewn on the beach. Beneath the surface, comb jellies often abound, as do siphonophores, salps, and scyphomedusae.

The pelagic environment was once thought to present few distinct habitats, in contrast to the array of niches within the benthic environment. Because of its apparent uniformity, the pelagic realm was understood to be distinguished simply by plankton of different sizes. Small-scale variations in the pelagic environment, however, have been discovered that affect biotic distributions. Living and dead matter form organic aggregates called marine snow to which members of the plankton community may adhere, producing patchiness in biotic distributions. Marine snow includes structures such as aggregates of cells and mucus as well as drifting macroalgae and other flotsam that range in size from 0.5 millimetre to 1 centimetre (although these aggregates can be as small as 0.05 millimetre and as large as 100 centimetres). Many types of microbes, phytoplankton, and zooplankton stick to marine snow, and some grazing copepods and predators will feed from the surface of these structures. Marine snow is extremely abundant at times, particularly after plankton blooms. Significant quantities of organic material from upper layers of the ocean may sink to the ocean floor as marine snow, providing an important source of food for bottom dwellers. Other structures that plankton respond to in the marine environment include aggregates of phytoplankton cells that form large rafts in tropical and temperate waters of the world (e.g., cells of Oscillatoria [Trichodesmium] erthraeus) and various types of seaweed (e.g., Sargassum, Phyllospora, Macrocystis) that detach from the seafloor and drift.

Nekton

Nekton are the active swimmers of the oceans and is often the best-known organisms of marine waters. Nekton is the top predators in most marine food chains. The distinction between nekton and plankton is not always sharp. As mentioned above, many large marine animals, such as marlin and tuna, spend the larval stage of their lives as plankton and their adult stage as large and active members of the nekton. Other organisms such as krill are referred to as both micronekton and macrozooplankton.

The vast majority of nekton are vertebrates (e.g., fishes, reptiles, and mammals), mollusks, and crustaceans. The most numerous group of nekton are the fishes, with approximately 16,000 species. Nekton are found at all depths and latitudes of marine waters. Whales, penguins, seals, and icefish abound in polar waters. Lantern fish (family Myctophidae) are common in the aphotic zone along with gulpers (Saccopharynx), whalefish (family Cetomimidae), seven-gilled sharks, and others. Nekton diversity is greatest in tropical waters, where in particular there are large numbers of fish species.

The largest animals on the Earth, the blue whales (Balaenoptera musculus), which grow to 25 to 30 metres long, are members of the nekton. These huge mammals and other baleen whales (order Mysticeti), which are distinguished by fine filtering plates in their mouths, feed on plankton and micronekton as do whale sharks (Rhinocodon typus), the largest fish in the world (usually 12 to 14 metres long, with some reaching 17 metres). The largest carnivores that consume large prey include the toothed whales (order Odontoceti—for example, the killer whales, Orcinus orca), great white sharks (Carcharodon carcharias), tiger sharks (Galeocerdo cuvier), black marlin (Makaira indica), bluefin tuna (Thunnus thynnus), and giant groupers (Epinephelus lanceolatus).

Nekton form the basis of important fisheries around the world. Vast schools of small anchovies, herring, and sardines generally account for one-quarter to one-third of the annual harvest from the ocean. Squid are also economically valuable nekton. Halibut, sole, and cod are demersal (i.e., bottom-dwelling) fish that are commercially important as food for humans. They are generally caught in continental shelf waters. Because pelagic nekton often abound in areas of upwelling where the waters are nutrient-rich, these regions also are major fishing areas.

Benthos

Organisms are abundant in surface sediments of the continental shelf and in deeper waters, with a great diversity found in or on sediments. In shallow waters, beds of seagrass provide a rich habitat for polychaete worms, crustaceans (e.g., amphipods), and fishes. On the surface of and within intertidal sediments most animal activities are influenced strongly by the state of the tide. On many sediments in the photic zone, however, the only photosynthetic organisms are microscopic benthic diatoms.

Benthic organisms can be classified according to size. The macrobenthos are those organisms larger than 1 millimetre. Those that eat organic material in sediments are called deposit feeders (e.g., holothurians, echinoids, gastropods), those that feed on the plankton above are the suspension feeders (e.g., bivalves, ophiuroids, crinoids), and those that consume other fauna in the benthic assemblage are predators (e.g., starfish, gastropods). Organisms between 0.1 and 1 millimetre constitute the meiobenthos. These larger microbes, which include foraminiferans, turbellarians, and polychaetes, frequently dominate benthic food chains, filling the roles of nutrient recycler, decomposer, primary producer, and predator. The microbenthos are those organisms smaller than 1 millimetre; they include diatoms, bacteria, and ciliates.

Organic matter is decomposed aerobically by bacteria near the surface of the sediment where oxygen is abundant. The consumption of oxygen at this level, however, deprives deeper layers of oxygen, and marine sediments below the surface layer are anaerobic. The thickness of the oxygenated layer varies according to grain size, which determines how permeable the sediment is to oxygen and the amount of organic matter it contains. As oxygen concentration diminishes, anaerobic processes come to dominate. The transition layer between oxygen-rich and oxygen-poor layers is called the redox discontinuity layer and appears as a gray layer above the black anaerobic layers. Organisms have evolved various ways of coping with the lack of oxygen. Some anaerobes release hydrogen sulfide, ammonia, and other toxic reduced ions through metabolic processes. The thiobiota, made up primarily of microorganisms, metabolize sulfur. Most organisms that live below the redox layer, however, have to create an aerobic environment for themselves. Burrowing animals generate a respiratory current along their burrow systems to oxygenate their dwelling places; the influx of oxygen must be constantly maintained because the surrounding anoxic layer quickly depletes the burrow of oxygen. Many bivalves (e.g., Mya arenaria) extend long siphons upward into oxygenated waters near the surface so that they can respire and feed while remaining sheltered from predation deep in the sediment. Many large mollusks use a muscular "foot" to dig with, and in some cases they use it to propel themselves away from predators such as starfish. The consequent "irrigation" of burrow systems can create oxygen and nutrient fluxes that stimulate the production of benthic producers (e.g., diatoms).

Not all benthic organisms live within the sediment; certain benthic assemblages live on a rocky substrate. Various phyla of algae—Rhodophyta (red), Chlorophyta (green), and Phaeophyta (brown)—are abundant and diverse in the photic zone on rocky substrata and are important producers. In intertidal regions algae are most abundant and largest near the low-tide mark. Ephemeral algae such as Ulva, Enteromorpha, and coralline algae cover a broad range of the intertidal. The mix of algae species found in any particular locale is dependent on latitude and also varies greatly according to wave exposure and the activity of grazers.

For example, Ascophyllum spores cannot attach to rock in even a gentle ocean surge; as a result this plant is largely restricted to sheltered shores. The fastest-growing plant—adding as much as 1 metre per day to its length—is the giant kelp, Macrocystis pyrifera, which is found on subtidal rocky reefs. These plants, which may exceed 30 metres in length, characterize benthic habitats on many temperate reefs. Large laminarian and fucoid algae are also common on temperate rocky reefs, along with the encrusting (e.g., Lithothamnion) or short tufting forms (e.g., Pterocladia). Many algae on rocky reefs are harvested for food, fertilizer, and pharmaceuticals. Macroalgae are relatively rare on tropical reefs where corals abound, but Sargassum and a diverse assemblage of short filamentous and tufting algae are found, especially at the reef crest. Sessile and slow-moving invertebrates are common on reefs. In the intertidal and subtidal regions herbivorous gastropods and urchins abound and can have a great influence on the distribution of algae. Barnacles are common sessile animals in the intertidal. In the subtidal regions, sponges, ascidians, urchins, and anemones are particularly common where light levels drop and current speeds are high. Sessile assemblages of animals are often rich and diverse in caves and under boulders.

Reef-building coral polyps (Scleractinia) are organisms of the phylum Cnidaria that create a calcareous substrate upon which a diverse array of organisms live. Approximately 700 species of corals are found in the Pacific and Indian oceans and belong to genera such as Porites, Acropora, and Montipora. Some of the world's most complex ecosystems are found on coral reefs. Zooxanthellae are the photosynthetic, single-celled algae that live symbiotically within the tissue of corals and help to build the solid calcium carbonate matrix of the reef. Reef-building corals are found only in waters warmer than 18 °C; warm temperatures are necessary, along with high light intensity, for the coral-algae complex to secrete calcium carbonate. Many tropical islands are composed entirely of hundreds of metres of coral built atop volcanic rock.

Links between the Pelagic Environments and the Benthos

Considering the pelagic and benthic environments in isolation from each other should be done cautiously because the two are interlinked in many ways. For example, pelagic plankton are an important source of food for animals on soft or rocky bottoms. Suspension feeders such as anemones and barnacles filter living and dead particles from the surrounding water while detritus feeders graze on the accumulation of particulate material raining from the water column above. The molts of crustaceans, plankton feces, dead plankton, and marine snow all contribute to this rain of fallout from the pelagic environment to the ocean bottom. This fallout can be so intense in certain weather patterns—such as the El Niño condition—that benthic animals on soft bottoms are smothered and die. There also is variation in the rate of fallout of the plankton according to seasonal cycles of production. This variation can create seasonality in the abiotic zone where there is little or no variation in temperature or light. Plankton form marine sediments, and many types of fossilized protistan plankton, such as foraminiferans and coccoliths, are used to determine the age and origin of rocks.

Organisms of the Deep-sea Vents

Producers were discovered in the aphotic zone when exploration of the deep sea by submarine became common in the 1970s. Deep-sea hydrothermal vents now are known to be relatively common in areas of tectonic activity (e.g., spreading ridges). The vents are a non photosynthetic source of organic carbon available to organisms. A diversity of deep-sea organisms including mussels, large bivalve clams, and vestimentiferan worms are supported by bacteria that oxidize sulfur (sulfide) and derive chemical energy from the reaction. These organisms are referred to as chemoautotrophic, or chemosynthetic, as opposed to photosynthetic, organisms. Many of the species in the vent fauna have developed symbiotic relationships with chemoautotrophic bacteria, and as a consequence the megafauna are principally responsible for the primary production in the vent assemblage. The situation is analogous to that found on coral reefs where individual coral polyps have symbiotic relationships with zooxanthellae. In addition to symbiotic bacteria there is a rich assemblage of free-living bacteria around vents. For example, Beggiatoas-like bacteria often form conspicuous weblike mats on any hard surface; these mats have been shown to have chemoautotrophic metabolism. Large numbers of brachyuran (e.g., Bythograea) and galatheid crabs, large sea anemones (e.g., Actinostola callasi), copepods, other plankton, and some fish—especially the eelpout Thermarces Cerberus are found in association with vents.

Hydrothmussels
Galatheid crabs and shrimp grazing on the bacterial filaments that grows on the shells of the hydrothermal mussels covering the Northwest Eifuku volcano in the Mariana Arc region

Patterns and Processes Influencing the Structure of Marine Assemblages

Distribution and Dispersal

The distribution patterns of marine organisms are influenced by physical and biological processes in both ecological time (tens of years) and geologic time (hundreds to millions of years). The shapes of the Earth's oceans have been influenced by plate tectonics, and as a consequence the distributions of fossil and extant marine organisms also have been affected. Vicariance theory argues that plate tectonics has a major role in determining biogeographic patterns. For example, Australia was once—90 million years ago—close to the South Pole and had few coral reefs. Since then Australia has been moving a few millimetres each year closer to the Equator. As a result of this movement and local oceanographic conditions, coral reef environments are extending ever so slowly southward. Dispersal may also have an important role in biogeographic patterns of abundance. The importance of dispersal varies greatly with local oceanographic features, such as the direction and intensity of currents and the biology of the organisms. Humans can also have an impact on patterns of distribution and the extinction of marine organisms. For example, fishing intensity in the Irish Sea was based on catch limits set for cod with no regard for the biology of other species. One consequence of this practice was that the local skate, which had a slow reproductive rate, was quickly fished to extinction.

A characteristic of many marine organisms is a bipartite life cycle, which can affect the dispersal of an organism. Most animals found on soft and hard substrata, such as lobsters, crabs, barnacles, fish, polychaete worms, and sea urchins, spend their larval phase in the plankton and in this phase are dispersed most widely. The length of the larval phase, which can vary from a few minutes to hundreds of days, has a major influence on dispersal. For example, wrasses of the genus Thalassoma have a long larval life, compared with many other types of reef fish, and populations of these fish are well dispersed to the reefs of isolated volcanic islands around the Pacific. The bipartite life cycle of algae also affects their dispersal, which occurs through algal spores. Although in general, spores disperse only a short distance from adult plants, limited swimming abilities—Macrocystis spores have flagella—and storms can disperse spores over greater distances.

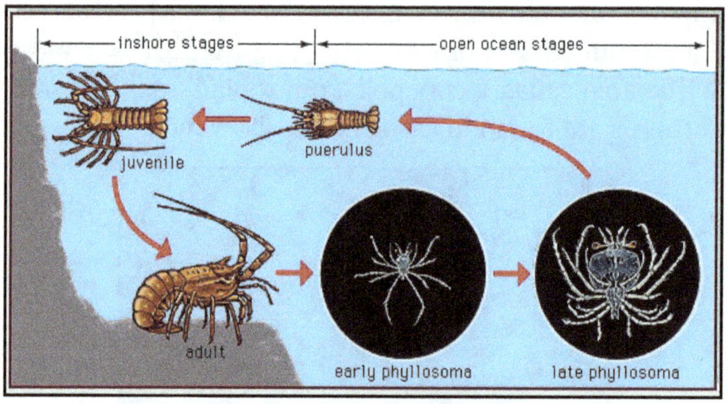

Life cycle of a palinurid lobster

Migrations of Marine Organisms

The migrations of plankton and nekton throughout the water column in many parts of the world are well described. Diurnal vertical migrations are common. For example, some types of plankton, fish, and squid remain beneath the photic zone during the day, moving toward the surface after dusk and returning to the depths before dawn. It is generally argued that marine organisms migrate in response to light levels. This behaviour may be advantageous because by spending the daylight hours in the dim light or darkness beneath the photic zone plankton can avoid predators that locate their prey visually. After the Sun has set, plankton can raise to the surface waters where food is more abundant and where they can feed safely under the cover of darkness.

Larval forms can facilitate their horizontal transport along different currents by migrating vertically. This is possible because currents can differ in direction according to depth (e.g., above and below haloclines and thermoclines), as is the case in estuaries.

In coastal waters many larger invertebrates (e.g., mysids, amphipods, and polychaete worms) leave the cover of algae and sediments to migrate into the water column at night. It is thought that these animals disperse to different habitats or find mates by swimming when visual predators find it hard to see them. In some cases only one sex will emerge at night, and often that sex is morphologically better suited for swimming.

Horizontal migrations of fish that span distances of hundreds of metres to tens of kilometres are common and generally related to patterns of feeding or reproduction. Tropical coral trout (Plectropomus species) remain dispersed over a reef for most of the year, but adults will aggregate at certain locations at the time of spawning. Transoceanic migrations (greater than 1,000 kilometres) are observed in a number of marine vertebrates, and these movements often relate to requirements of feeding and reproduction. Bluefin tuna (Thunnus thynnus) traverse the Atlantic Ocean in a single year; they spawn in the Caribbean, then swim to high latitudes of the Atlantic to feed on the rich supply of fish. Turtles and sharks also migrate great distances.

Fish that spend their lives in both marine and freshwater systems (diadromous animals) exhibit some of the most spectacular migratory behaviour. Anadromous fishes (those that spend most of their lives in the sea but migrate to fresh water to spawn) such as Atlantic salmon (Salmo salar) also have unique migratory patterns. After spawning, the adults die. Newly hatched fish (alevin) emerge from spawned eggs and develop into young fry that move down rivers toward the sea. Juveniles (parr) grow into larger fish (smolt) that convene near the ocean. When the adult fish are ready to spawn, they return to the river in which they were born (natal river), using a variety of environmental cues, including the Earth's magnetic field, the Sun, and water chemistry. It is believed that the thyroid gland has a role in imprinting the water chemistry of the natal river on the fish. Freshwater eels such as the European eel (Anguilla anguilla) undertake great migrations from fresh water to spawn in the marine waters of the Sargasso Sea (catadromous migrations), where they die. Eel larvae, called leptocephalus larvae, drift back to Europe in the Gulf Stream.

Dynamics of Populations and Assemblages

A wide variety of processes influence the dynamics of marine populations of individual species and the composition of assemblages (e.g., collections of populations of different species that live in the same area). With the exception of marine mammals such as whales, fish that bear live young (e.g., embiotocid fish), and brooders (i.e., fauna that incubate their offspring until they emerge as larvae or juveniles), most marine organisms produce a large number of offspring of which few survive. Processes that affect the plankton can have a great influence on the numbers of young that survive to be recruited, or relocated, into adult populations. The survival of larvae may depend on the abundance of food at various times and in various places, the number of predators, and oceanographic features that retain larvae near suitable nursery areas. The number of organisms recruited to benthic and pelagic systems may ultimately determine the size of adult populations and therefore the relative abundance of species in marine assemblages. However, many processes can affect the survival of organisms after recruitment. Predators eat recruits, and mortality rates in prey species can vary with time and space, thus changing original population patterns established in recruitment.

Patterns of colonization and succession can have a significant impact on benthic assemblages. For example, when intertidal reefs are cleared experimentally, the assemblage of organisms that colonize the bare space often reflects the types of larvae available in local waters at the time. Tube worms may dominate if they establish themselves first; if they fail to do so, algal spores may colonize the shore first and inhibit the settlement of these worms. Competition between organisms may also play a role. Long-term data gathered over periods of more than 25 years from coral reefs have demonstrated that some corals (e.g., Acropora cytherea) competitively overgrow neighbouring corals. Physical disturbance from hurricanes destroys many corals, and during regrowth competitively inferior species can coexist with normally dominant species on the reef. Chemical defenses of sessile organisms also can deter the growth or cause increased mortality of organisms that settle on them. Ascidian larvae (e.g., Podoclavella) often avoid settling on sponges (e.g., Mycale); when this does occur, the larvae rarely reach adulthood.

Although the processes that determine species assemblages may be understood, variations occur in the composition of the plankton that make it difficult to predict patterns of colonization with great accuracy.

Biological Productivity

Primary productivity is the rate at which energy is converted by photosynthetic and chemosynthetic autotrophs to organic substances. The total amount of productivity in a region or system is gross primary productivity. A certain amount of organic material is used to sustain the life of producers; what remains is net productivity. Net marine primary productivity is the amount of organic material available to support the consumers (herbivores and carnivores) of the sea. The standing crop is the total biomass (weight) of vegetation. Most primary productivity is carried out by pelagic phytoplankton, not benthic plants.

Most primary producers require nitrogen and phosphorus, which are available in the ocean as nitrate, nitrite, ammonia, and phosphorus. The abundances of these molecules and the intensity and quality of light exert a major influence on rates of production. The two principal categories of producers (autotrophs) in the sea are pelagic phytoplankton and benthic microalgae and macroalgae. Benthic plants grow only on the fringe of the world's oceans and are estimated to produce only 5 to 10 percent of the total marine plant material in a year. Chemoautotrophs are the producers of the deep-sea vents.

Primary productivity is usually determined by measuring the uptake of carbon dioxide or the output of oxygen. Production rates are usually expressed as grams of organic carbon per unit area per unit time. The productivity of the entire ocean is estimated to be approximately 16×10^{10} tons of carbon per year, which is about eight times that of the land.

Pelagic Food Chain

Food chains in coastal waters of the world are generally regulated by nutrient concentrations. These concentrations determine the abundance of phytoplankton, which in turn provide food for the primary consumers, such as protozoa and zooplankton, which the higher-level consumers—fish, squid, and marine mammals—prey upon. It had been thought that phytoplankton in the 5- to 100-micrometre size range were responsible for most of the primary production in the sea and that grazers such as copepods controlled the numbers of phytoplankton. Data gathered since 1975, however, indicate that the system is much more complex than this. It is now thought that most primary production in marine waters of the world is accomplished by single-celled 0.5- to 10-micrometre phototrophs (bacteria and protists). Moreover, heterotrophic protists (phagotrophic protists) are now viewed as the dominant controllers of both bacteria and primary production in the sea. Current models of pelagic marine food chains picture complex interactions within a microbial food web. Larger metazoans are supported by the production of autotrophic and heterotrophic cells.

Upwelling

The most productive waters of the world are in regions of upwelling. Upwelling in coastal waters brings nutrients toward the surface. Phytoplankton reproduce rapidly in these conditions, and grazing zooplankton also multiply and provide abundant food supplies for nekton. Some of the world's richest fisheries are found in regions of upwelling—for example, the temperate waters off

Peru and California. If upwelling fails, the effects on animals that depend on it can be disastrous. Fisheries also suffer at these times, as evidenced by the collapse of the Peruvian anchovy industry in the 1970s. The intensity and location of upwelling are influenced by changes in atmospheric circulation, as exemplified by the influence of El Niño conditions.

Seasonal Cycles of Production

Cycles of plankton production vary at different latitudes because seasonal patterns of light and temperature vary dramatically with latitude. In the extreme conditions at the poles, plankton populations crash during the constant darkness of winter and bloom in summer with long hours of light and the retreat of the ice field. In tropical waters, variation in sunlight and temperature is slight, nutrients are present in low concentrations, and planktonic assemblages do not undergo large fluctuations in abundance. There are, however, rapid cycles of reproduction and high rates of grazing and predation that result in a rapid turnover of plankton and a low standing crop. In temperate regions plankton abundance peaks in spring as temperature and the length and intensity of daylight increase. Moreover, seasonal winter storms usually mix the water column, creating a more even distribution of the nutrients, which facilitates the growth of phytoplankton. Peak zooplankton production generally lags behind that of phytoplankton, while the consumption of phytoplankton by zooplankton and phagotrophic protists is thought to reduce phytoplankton abundance. Secondary peaks in abundance occur in autumn. Seasonal peaks of some plankton are very conspicuous, and the composition of the plankton varies considerably. In spring and early summer many fish and invertebrates spawn and release eggs and larvae into the plankton, and, as a result, the meroplanktonic component of the plankton is higher at these times. General patterns of plankton abundance may be further influenced by local conditions. Heavy rainfall in coastal regions (especially areas in which monsoons prevail) can result in nutrient-rich turbid plumes (i.e., estuarine or riverine plumes) that extend into waters of the continental shelf. Changes in production, therefore, may depend on the season, the proximity to fresh water, and the timing and location of upwelling, currents, and patterns of reproduction.

Coral Reef Ecosystem

Coral reef is the term for any fully marine, underwater ecosystem with a calcium carbonate foundation produced by the skeletons of living organisms, a substantial number of which are corals. Generally, coral reefs are considered to be those limestone formations and associated biotic communities in which the corals have a symbiotic relationship with algae and thus they are found in shallow and warm marine waters with significant sunlight penetration. However, there also are corals that do not have associated algae and these can produce reef formations in deep, dark, and cold marine environments. These deep-water reefs are called coral bioherms, which some consider as a type of coral reef, and others consider as distinct from the term coral reef. The calcium carbonate found in coral reefs not only comes from skeletal material built up by coral over time, but also from such living organisms as coralline algae and foraminifera's, among others.

The calcareous structure produced by the living organisms makes for an excellent habitat for animal and plant life. Known as "rainforests of the sea" and "oases in the desert of the ocean," coral reefs are among the most rich and diverse ecosystems on Earth. Despite covering less than one tenth of one percent of the world's oceans, coral reefs provide a habitat for twenty-five percent of all marine species, resulting in complex food webs. Paradoxically, these biologically rich systems flourish even though they are surrounded by ocean waters that provide few nutrients.

In addition to their vital ecological function, coral reefs provide important economic values for human beings, delivering ecosystem services to tourism, fisheries, and shoreline protection. The annual global economic value of coral reefs has been estimated at $US 375 billion. However, coral reefs are fragile ecosystems, and despite these values, various anthropogenic factors have resulted in the loss of substantial amounts of coral reefs in recent decades. There are estimates that 75% of the world's coral reefs are presently threatened as a result of human activities. In recent years, people have devoted a greater attention to protection of coral reefs, including efforts to effectively manage them in protected marine parks and reserves.

The foundation for the underwater structures of coral reefs comes from the calcium carbonate secreted by corals and other living organisms. Corals are those marine invertebrates of the phylum Cnidaria and the class Anthozoa that have external or internal calcareous skeletons; the skeletons of these animals are also called coral. Corals exist as small polyps, typically in colonies of many individuals and commonly attached to a hard surface. The polyps secrete a hard carbonate exoskeleton which provides support and protection for the body of each polyp. New corals grown on the foundation of the skeletal material left by earlier reef-building corals, thus extending the coral reef, with the living reef community on the surface of the reef. Most coral reefs are built from stony corals. Other organisms that deposit calcium carbonate include various algae, foraminiferans, mollusks, annelids, and echinoderms.

Corals are found in marine waters that contain few nutrients. Although corals can catch plankton, these animals obtain much of their nutrients from symbiotic unicellular dinoflagellates (type of photosynthetic algae) called zooxanthellae. Consequently, most corals are dependent upon ample sunlight and for that reason are usually found not far beneath the surface, restricted to the euphotic zone typically above about 70 meters from the surface. The reefs from these corals are found in tropical and subtropical marine waters, where the water temperature does not fall below 18 °C (64 °F) for extended amounts of time, and generally is above 22 °C (72 °F), although corals live outside this range. Coral reefs formed from such symbiotic corals are almost exclusively limited to the range of latitudes from 30 °N to 30 °S,.

The most extensive development of extant coral reef is the Great Barrier Reef off the coast of Queensland, Australia. Indonesia is home to almost 600 of the world's approximately 800 known coral reef-building coral species.

However, some corals, notably the genus Lophelia, do not have associated algae, and can live in much deeper and colder water. These form reefs as well, although these are typically called "coral bioherms." Some authorities consider the term coral reefs to include these deep-water coral bioherms, based on their biological and physical characteristics. However, other authorities distinguish these terms. Cold-water corals can be found in such locales as off the coast of Norway (north to at least 69 °14.24' N) and the Darwin Mounds off western Scotland. Deep-water coral bioherms are quite common off the southeastern United States.

The term coral reef does not refer only the physical structure; it is an ecosystem. An ecosystem consists both of the abiotic environment as well as the biotic community (an assemblage of plant, animal, and other living organisms). Coral reefs form some of the most diverse ecosystems on Earth, and have been called "rainforests of the sea." They occupy less than one tenth of one percent of the world ocean surface, about half the area of France, yet they provide a home for twenty-five percent of all marine species, including fish, mollusks, worms, crustaceans, echinoderms, sponges, tunicates and other cnidarians.

Coral reefs are fragile ecosystems, partly because they are very sensitive to water temperature. They are under threat from climate change, ocean acidification, blast fishing, cyanide fishing for aquarium fish, overuse of reef resources, and harmful land-use practices, including urban and agricultural runoff and water pollution, which can harm reefs by encouraging excess algae growth.

Close up of polyps arrayed on a coral, waving their tentacles: There can be thousands of polyps on a single coral branch

Coral Biology

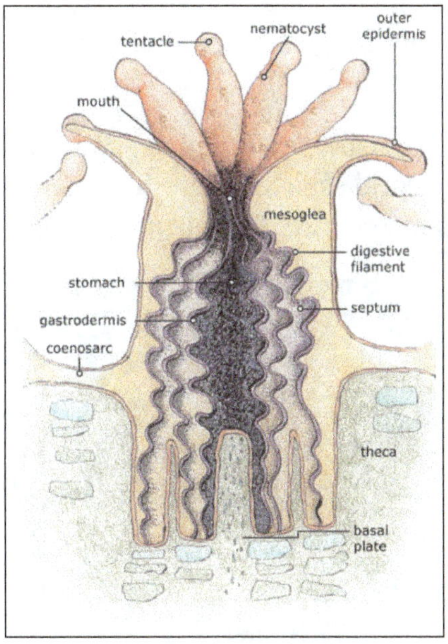

Anatomy of a coral polyp

Live coral are small marine invertebrates (phylum Cnidaria, class Anthozoa) embedded in calcium carbonate shells. As members of the Anthozoa class ("flower animals,"), they live only as the vase-shaped polyps, characterized by a tubular body with tentacles around the mouth, and they are mostly sedentary after the larval stage. Corals typically live in colonies of many individuals and commonly attached to a hard surface, giving the appearance of plants or rocks. Polyps are usually tiny, but they can range in size from a pinhead to a foot across.

These accumulations of polyps can be arranged in a wide diversity of shapes. For example, branching corals have branches and these branches have further branches; table corals create table-like structures with the branches fused; and digitate corals give the appearance of fingers. Other colony structures include elkhorn corals or staghorn corals, which have flattened and large branches; massive corals, which are shaped like balls or boulders and can be the size of eggs or houses; submassive corals or pillar corals, which look like columns or wedges protruding from a base; mushroom corals that resemble the tops of mushrooms; and cup corals, which look like eggs cups or like cups that have been elongated, twisted, or squashed.

Reef-building or hermatypic corals live only in the euphotic zone, the depth to which sufficient sunlight penetrates the water for photosynthesis to occur. Coral polyps do not themselves photosynthesize, but have a symbiotic relationship with single-celled organisms called zooxanthellae; these organisms live within the tissues of polyps and provide organic nutrients that nourish the polyp. Because of this relationship, coral reefs grow much faster in clear water, which admits more sunlight. Indeed, the relationship is responsible for coral reefs in the sense that without their symbionts, coral growth would be too slow for the corals to form significant reef structures. Corals get up to 90% of their nutrients from their zooxanthellae symbionts.

Reefs grow as the coral polyps and other organisms deposit calcium carbonate, the basis of coral, as a skeletal structure beneath and around themselves, pushing the coral head's top upwards and outwards. Many other organisms living in the reef community contribute skeletal calcium carbonate in the same manner. Coralline algae are important contributors to reef structure in those parts of the reef subjected to the greatest forces by waves (such as the reef front facing the open ocean). These algae deposit limestone in sheets over the reef surface, thereby strengthening it.

Waves, grazing fish (such as parrotfish), sea urchins, sponges, and other forces and organisms act as bioeroders, breaking down coral skeletons into fragments that settle into spaces in the reef structure or form sandy bottoms in associated reef lagoons.

Brain coral

Table coral

Staghorn coral

Spiral wire coral

Pillar coral

Reef Structure

Types

The three principal reef types are:

- Fringing reef: The most common type, a fringing reef is one that is directly attached to a shore or borders it with an intervening shallow channel or lagoon. Many fringing reefs can be found in the Florida Keys.
- Barrier reef: A barrier reef also borders the shoreline, but it is separated from a mainland or island shore by a deep channel or lagoon; that is, it is separated by a greater distance from the shoreline (a mile or more) and there is open, often deep water separating it from the adjacent land mass. The Great Barrier Reef around Australia is the world's largest barrier reef.
- Atoll reef: An atoll reef is a more or less continuous circular or oval barrier reef extending all the way around a lagoon without a central island. In other words, if a fringing reef or barrier reef were to form around a volcanic island, and then the island were to sink entirely below the sea level, while the atoll continues to expand upward, an atoll would be formed. Passages may exist between the central lagoon and the ocean. An example is the Bikini Atoll.

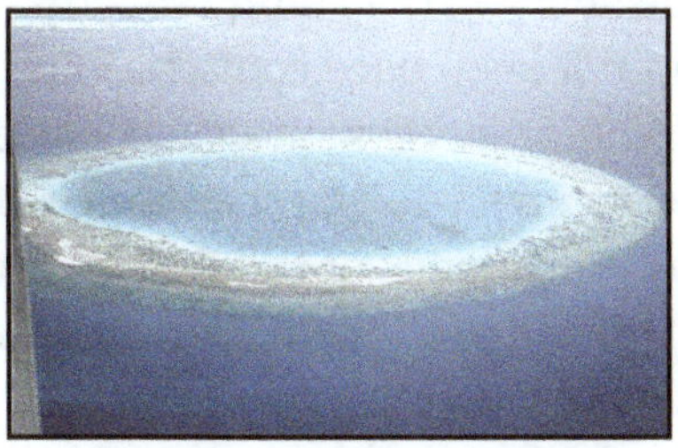

A small atoll in the Maldives

Inhabited cay in the Maldives

Other reef types or variants are:

- Patch reef – An isolated, comparatively small reef outcrop, usually within a lagoon or embayment, often circular and surrounded by sand or seagrass. Patch reefs, which are basically small patches of reef, are common.
- Apron reef – A short reef resembling a fringing reef, but more sloped; extending out and downward from a point or peninsular shore.

- Bank reef or platform reef – Isolated reefs with a linear or semi-circular shaped-outline and flat top, larger than a patch reef.
- Ribbon reef – A long, narrow, possibly winding or sinuous reef, usually associated with an atoll lagoon and parallel the shelf break.
- Table reef – An isolated reef, approaching an atoll type, but without a lagoon.
- Habili – Reef in the Red Sea that does not reach the surface near enough to cause visible surf, although it may a hazard to ships (from the Arabic for "unborn").
- Microatolls – Certain species of corals form communities called microatolls. The vertical growth of microatolls is limited by average tidal height. By analyzing growth morphologies, microatolls offer a low resolution record of patterns of sea level change. Fossilized microatolls can also be dated using radioactive carbon dating. Such methods have been used to reconstruct Holocene sea levels.
- Cays – Small, low-elevation, sandy islands formed on the surface of a coral reef. Material eroded from the reef piles up on parts of the reef or lagoon, forming an area above sea level. Plants can stabilize cays enough to become habitable by humans. Cays occur in tropical environments throughout the Pacific, Atlantic, and Indian Oceans (including the Caribbean and on the Great Barrier Reef and Belize Barrier Reef), where they provide habitable and agricultural land for hundreds of thousands of people.
- When a coral reef cannot keep up with the sinking of a volcanic island, a seamount or guyot is formed. The tops of seamounts and guyots are below the surface. Seamounts are rounded at the top and guyots are flat. The flat top of the guyot, also called a tablemount, is due to erosion by waves, winds, and atmospheric processes.

Formation

Most existing coral reefs were formed after the last glacial period when melting ice caused the sea level to rise and flood the continental shelves. This means that most coral reefs are less than 10,000 years old. As coral reef communities were established on the shelves, they built reefs that grew upwards, keeping pace with the rise in sea level. Reefs that did not keep pace could become drowned reefs, covered by so much water that there was insufficient light for further survival. Charles Darwin set out his theory of the formation of atoll reefs, an idea he conceived during the voyage of the Beagle. His theory was that atolls were formed by the uplift and subsidence of the Earth's crust under the oceans. Darwin's theory sets out a sequence of three stages in atoll formation. It starts with a fringing reef forming around an extinct volcanic island as the island and ocean floor subsides. As the subsidence continues, the fringing reef becomes a barrier reef, and ultimately an atoll reef.

Darwin's theory starts with a volcanic island which becomes extinct

Ultimately the island sinks below the sea, and the barrier reef becomes an atoll enclosing an open lagoon

As the subsidence continues the fringing reef becomes a larger barrier reef further from the shore with a bigger and deeper lagoon inside

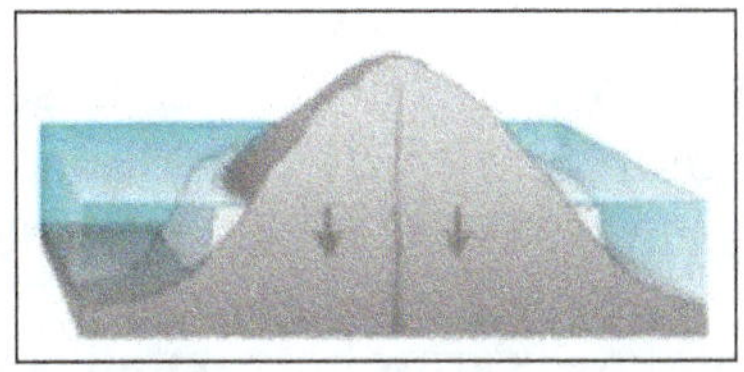

As the island and ocean floor subside, coral growth builds a fringing reef, often including a shallow lagoon between the land and the main reef

Darwin predicted that underneath each lagoon would be a bed rock base, the remains of the original volcano. Subsequent drilling proved this correct. Darwin's theory followed from his understanding that coral polyps thrive in the clean seas of the tropics where the water is agitated, but can only live within a limited depth of water, starting just below low tide. Where the level of the underlying land stays the same, the corals grow around the coast to form what he called fringing reefs, and can eventually grow out from the shore to become a barrier reef.

A fringing reef can take ten thousand years to form, and an atoll can take up to 30 million years

Where the land is rising, fringing reefs can grow around the coast, but coral raised above sea level dies and becomes white limestone. If the land subsides slowly, the fringing reefs keep pace by growing upwards on a base of dead coral, forming a barrier reef enclosing a lagoon between the reef and the land. A barrier reef can encircle an island, and once the island sinks below sea level a roughly circular atoll of growing coral continues to keep up with the sea level, forming a central lagoon. Barrier reefs and atolls do not usually form complete circles, but are broken in places by storms. Should the land subside too quickly or sea level rise too fast, the coral dies as it is below its habitable depth.

In general, the two main variables determining the geomorphology, or shape, of coral reefs are the nature of the underlying substrate on which they rest, and the history of the change in sea level relative to that substrate.

As an example of how coral reefs have formed on continental shelves, the current living reef structure of the Great Barrier Reef began growing about 20,000 years ago. Sea level was then 120 meters (around 400 feet) lower than it is today. As sea levels rose, the water and the corals encroached on what had been hills of the Australian coastal plain. By 13,000 years ago, the sea level had risen to 60 meters (around 200 feet) lower than at present, and the hills of the coastal plains were, by then, continental islands. As the sea level rise continued, water topped most of the continental islands. The corals could then overgrow the hills, forming the present cays and reefs. Sea level on the Great Barrier Reef has not changed significantly in the last 6,000 years, and the age of the present living reef structure is estimated to be between 6,000 and 8,000 years. Although the Great Barrier Reef formed along a continental shelf, and not around a volcanic island, Darwin's principles apply. The Great Barrier Reef development stopped at the barrier reef stage, since Australia is not about to submerge. It formed the world's largest barrier reef, 300-1000 meters (1000-3300 feet) from shore, stretching for 2000 kilometers (1200 miles).

Healthy coral reefs grow horizontally from 1 to 3 centimeters (0.4 - 1.2 inches) per year, and grow vertically anywhere from 1 to 25 centimeters (0.4 - 10 inches) per year; however, they grow only at depths shallow enough to permit ample sunlight, and cannot grow above sea level.

Ancient coral reefs

Beginning a few thousand years after hard skeletons were developed by marine organisms, coral reefs emerged. The times of maximum development were in the Middle Cambrian (513–501 Ma), Devonian (416–359 Ma), and Carboniferous (359–299 Ma), due to Order Rugosa extinct corals, and Late Cretaceous (100–65 Ma) and all Neogene (23 Ma–present), due to Order Scleractinia corals.

Not all reefs in the past were formed by corals: Early Cambrian (542–513 Ma) reefs resulted from calcareous algae and archaeocyathids (small animals with conical shape, probably related to sponges), while rudists (a type of bivalve) built Late Cretaceous (100–65 Ma) reefs.

Zones

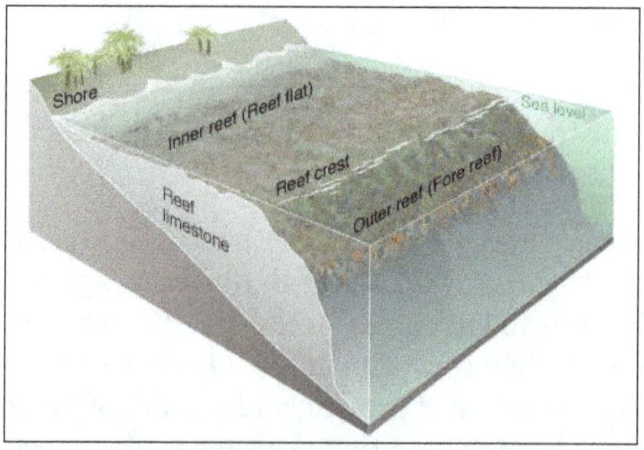

The three major zones of a coral reef: the fore reef, reef crest, and the back reef

Coral reef ecosystems contain distinct zones that represent different kinds of habitats. Usually three major zones are recognized: the fore reef, reef crest, and the back reef (frequently referred to as the reef lagoon). All three zones are physically and ecologically interconnected. Reef life and oceanic processes create opportunities for exchange of seawater, sediments, nutrients, and marine life among one another. Thus, they are integrated components of the coral reef ecosystem, each playing a role in the support of the reefs' abundant and diverse fish assemblages.

Most coral reefs exist in shallow waters less than fifty meters deep. Some inhabit tropical continental shelves where cool, nutrient rich upwelling does not occur, such as Great Barrier Reef. Others are found in the deep ocean surrounding islands or as atolls, such as in the Maldives. The reefs surrounding islands form when islands subside into the ocean and atolls form when an island subsides below the surface of the sea. Alternatively, Moyle and Cech distinguish six zones, though most reefs possess only some of the zones.

Water in the reef surface zone is often agitated

This diagram represents a reef on a continental shelf. The water waves at the left travel over the off-reef floor until they encounter the reef slope or fore reef. Then the waves pass over the shallow reef crest. When a wave enters shallow water it shoals, that is, it slows down and the wave height increases.

- The reef surface is the shallowest part of the reef: It is subject to the surge and the rise and fall of tides. When waves pass over shallow areas, they shoal, as shown in the diagram at the right. This means that the water is often agitated. These are the precise condition under which coral flourish. Shallowness means there is plenty of light for photosynthesis by the symbiotic zooxanthellae, and agitated water promotes the ability of coral to feed on plankton. However other organisms must be able to withstand the robust conditions to flourish in this zone.
- The off-reef floor is the shallow sea floor surrounding a reef: This zone occurs by reefs on continental shelves. Reefs around tropical islands and atolls drop abruptly to great depths, and do not have a floor. Usually sandy, the floor often supports seagrass meadows which are important foraging areas for reef fish.
- The reef drop-off is, for its first 50 meters, habitat for many reef fish who find shelter on the cliff face and plankton in the water nearby. The drop-off zone applies mainly to the reefs surrounding oceanic islands and atolls.
- The reef face is the zone above the reef floor or the reef drop-off: "It is usually the richest habitat. Its complex growths of coral and calcareous algae provide cracks and crevices for protection, and the abundant invertebrates and epiphytic algae provide an ample source of food."
- The reef flat is a sandy bottomed flat area that can be behind the main reef, containing chunks of coral. "The reef flat may be a protective area bordering a lagoon, or it may be a flat, rocky area between the reef and the shore. In the former case, the number of fish species living in the area often is the highest of any reef zone."

- The reef lagoon is the structure when a coral reef completely encloses an area, thereby creating a quiet-water lagoon that usually contains small patches of reef."
- However, the "topography of coral reefs is constantly changing. Each reef is made up of irregular patches of algae, sessile invertebrates, and bare rock and sand. The size, shape and relative abundance of these patches changes from year to year in response to the various factors that favor one type of patch over another. Growing coral, for example, produces constant change in the fine structure of reefs. On a larger scale, tropical storms may knock out large sections of reef and cause boulders on sandy areas to move."

Distribution

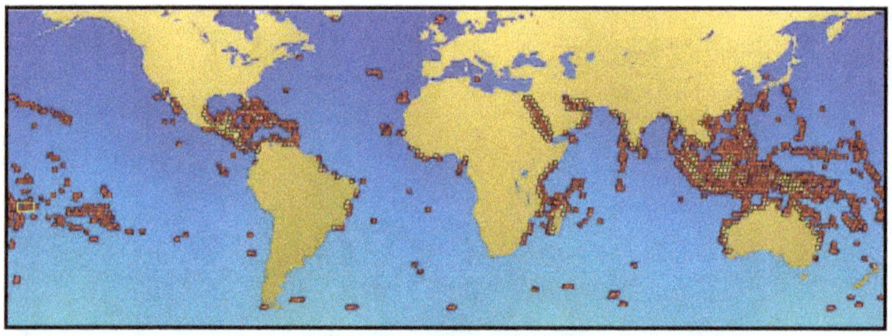

Locations of coral reefs

Coral reefs are estimated to cover 284,300 square kilometers (109,800 square miles), which is just under one tenth of one percent of the oceans' surface area. The Indo-Pacific region (including the Red Sea, Indian Ocean, Southeast Asia, and the Pacific) account for 91.9% of this total. Southeast Asia accounts for 32.3% of that figure, while the Pacific including Australia accounts for 40.8%. Atlantic and Caribbean coral reefs account for 7.6%.

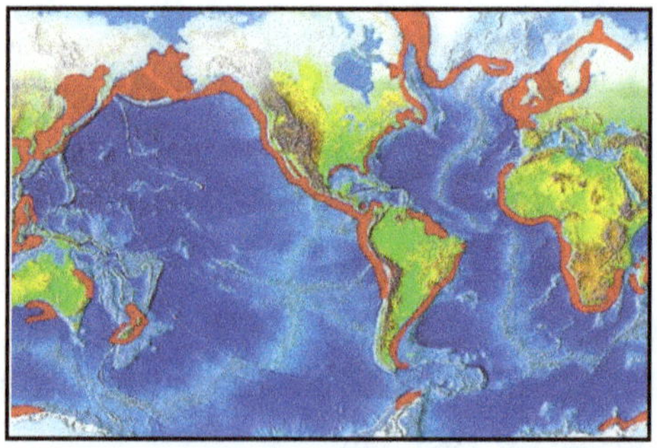

This map shows areas of upwelling in red. Coral reefs are not found in coastal areas where colder and nutrient-rich upwellings occur

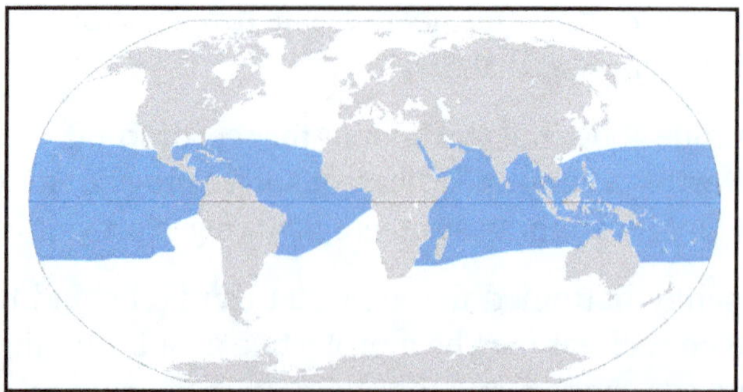

Boundary for 20 °C isotherms. Most corals live within this boundary. Note the cooler waters caused by upwelling on the south west coast of Africa and off the coast of Peru

Although corals exist both in temperate and tropical waters, shallow-water reefs form only in a zone extending from 30° N to 30° S of the equator. Tropical corals do not grow at depths of over 50 meters (about 160 feet). The optimum temperature for most coral reefs is 26-27 °C, and few reefs exist in waters below 18 °C (64 °F). However reefs in the Persian Gulf have adapted to temperatures of 13 °C in winter and 38 °C in summer. Deep water coral can exist at greater depths and colder temperatures. Although deep water corals can form reefs, very little is known about them.

Coral reefs are rare along the American west coast, as well as along the African west coast. This is due primarily to upwelling and strong cold coastal currents that reduce water temperatures in these areas (respectively the Peru, Benguela and Canary streams). Corals are seldom found along the coastline of South Asia from the eastern tip of India (Madras) to the border of Bangladesh and Myanmar. They are also rare along the coast around north-eastern South America and Bangladesh due to the freshwater release from the Amazon and Ganges Rivers respectively.

Principal Coral Reefs and Reef Areas

- The Great Barrier Reef - Largest, comprising over 2,900 individual reefs and 900 islands stretching for over 2600 kilometers (over 1600 miles), off Queensland, Australia.
- The Mesoamerican Barrier Reef System - Second largest, stretching 1000 kilometers (620 miles), from Isla Contoy at the tip of the Yucatán Peninsula down to the Bay Islands of Honduras.
- The New Caledonia Barrier Reef - Second longest double barrier reef, covering 1500 kilometers (930 miles).
- The Andros, Bahamas Barrier Reef - Third largest, following the east coast of Andros Island, Bahamas, between Andros and Nassau.
- The Red Sea - Includes 6000-year-old fringing reefs located around a 2000-kilometer (1240-mile) coastline.
- Pulley Ridge - Deepest photosynthetic coral reef, Florida.
- Numerous reefs scattered over the Maldives.
- Ghe Raja Ampat Islands in Indonesia's West Papua province offers the highest known marine diversity.

Ecology

Nutrient Paradox

During his voyage on the Beagle, Darwin described tropical coral reefs as oases in the desert of the ocean. He reflected on the paradox that tropical coral reefs, which are among the richest and most diverse ecosystems on earth, flourish surrounded by tropical ocean waters that provide hardly any nutrients.

Coral reefs cover less than one tenth of one percent of the surface of the world's ocean, yet they support over one-quarter of all marine species. This huge number of species results in complex food webs, with large predator fish eating smaller forage fish that eat yet smaller zooplankton and so on. However, all food webs eventually depend on plants, which are the primary producers. Coral reefs' primary productivity is very high, typically producting 5-10g C m^{-2} day^{-1} biomass.

One reason for the startling clarity of tropical waters is that they are deficient in nutrients and drifting plankton. Further, the sun shines year round in the tropics, warming the surface layer, making it less dense than subsurface layers. The warmer water is separated from the cooler water by a stable thermocline, where the temperature makes a rapid change. This keeps the warm surface waters floating above the cooler deeper waters. In most parts of the ocean, there is little exchange between these layers. Organisms that die in aquatic environments generally sink to the bottom where they decompose. This decomposition releases nutrients in the form of nitrogen (N), phosphorus, (P) and potassium (K). These nutrients are necessary for plant growth, but in the tropics they are not directly recycled back to the surface.

Plants form the base of the food chain, and need sunlight and nutrients to grow. In the ocean, these plants are mainly microscopic phytoplankton, which drift in the water column. They need sunlight for photosynthesis, which powers carbon fixation, so they are found only relatively near the surface. But they also need nutrients. Phytoplankton rapidly uses nutrients in the surface waters, and in the tropics these nutrients are not usually replaced because of the thermocline.

Retention and Recycling

Around coral reefs, lagoons fill in with material eroded from the reef and the island. They become havens for marine life, providing protection from waves and storms.

Most importantly, reefs recycle nutrients, which happens much less in the open ocean. In coral reefs and lagoons, producers include phytoplankton as well as seaweed and coralline algae, especially small types called turf algae, which pass nutrients to corals. The phytoplankton are eaten by fish and crustaceans, who also pass nutrients along the food web. Recycling ensures that fewer nutrients are needed overall to support the community.

Coral polyps

Coral reefs support many symbiotic relationships. In particular, zooxanthellae provides energy to coral in the form of glucose, glycerol, and amino acids. Zooxanthellae can provide up to 90% of a coral's energy requirements. In return, as an example of mutualism, the coral shelter the zooxanthellae, averaging one million for every cubic centimeter of coral, and provide a constant supply of the carbon dioxide it needs for photosynthesis.

The colour of corals depends on the type of zooxanthella they host

Corals also absorb nutrients, including inorganic nitrogen and phosphorus, directly from the water. Many corals extend their tentacles at night to catch zooplankton that brush them when the water is agitated. Zooplankton provides the polyp with nitrogen, and the polyp shares some of the nitrogen with the zooxanthellae, which also require this element. The varying pigments in different species of zooxanthellae give corals their different colors. Coral which loses its zooxanthellae becomes white and is said to be bleached, a condition that, unless corrected, can kill the coral.

Sponges are another key to explaining Darwin's paradox. They live in crevices in the coral reefs. They are efficient filter feeders, and in the Red Sea they consume about sixty percent of the phytoplankton that drifts by. The sponges eventually excrete nutrients in a form the corals can use.

Most coral polyps are nocturnal feeders. Here, in the dark, polyps have extended their tentacles to feed on zooplankton

The roughness of coral surfaces is the key to coral survival in agitated waters. Normally, a boundary layer of still water surrounds a submerged object, which acts as a barrier. Waves breaking on the extremely rough edges of corals disrupt the boundary layer, allowing the corals access to nutrients. Turbulent water thereby promotes rapid reef growth and lots of branching. Without the nutritional gains brought by rough coral surfaces, even the most effective recycling would leave corals wanting in nutrients. Cyanobacteria provide soluble nitrates for the reef via nitrogen fixation.

Coral reefs also often depend on surrounding habitats, such as seagrass meadows and mangrove forests, for nutrients. Seagrass and mangroves supply dead plants and animals that are rich in nitrogen and also serve to feed fish and animals from the reef by supplying wood and vegetation. Reefs in turn protect mangroves and seagrass from waves and produce sediment for the mangroves and seagrass to root in.

Biodiversity

Tube sponges attracting cardinal fishes, glassfishes, and wrasses

Reefs are home to a large variety of organisms, including fish, seabirds, sponges, Cnidarians (which includes some types of corals and jellyfish), worms, crustaceans (including shrimp, cleaner shrimp, spiny lobsters, and crabs), mollusks (including cephalopods), echinoderms (including starfish, sea urchins, and sea cucumbers), sea squirts, sea turtles, and sea snakes. Aside from humans, mammals are rare on coral reefs, with visiting cetaceans such as dolphins being the main exception. A few of these varied species feed directly on corals, while others graze on algae on the reef.

Fish

Coral reefs are home to a wide variety of tropical or reef fish, among them are the following:

- Fish that influence the coral (such as Labridae and parrotfish). These types of fish feed either on small animals living near the coral, seaweed, or on the coral itself. Fish that feed on small animals include cleaner fish (these fish feed on organisms that inhabit larger fish), bullet fish, and Balistidae (these eat sea urchins), while seaweed-eating fish include the Pomacentridae (damselfishes). Serranidae cultivate the seaweed by removing creatures feeding on it (such as sea urchins), and they remove inedible seaweeds. Fish that eat coral itself include parrotfish and butterflyfish.

- Fish that cruise the boundaries of the reef or nearby seagrass meadows. These include predatory fish such as pompanos, groupers, Horse mackerels, certain types of shark, Epinephelus marginatus, barracudas, and snappers). Herbivorous and plankton-eating fish also populate reefs. Seagrass-eating fish include Horse mackerel, snapper, Pagellus, and Conodon. Plankton-eating fish include Caesio, manta ray, chromis, Holocentridae, and pterapogon kauderni.

Organisms can cover every square inch of a coral reef

Fish that swim in coral reefs can be as colorful as the reef. Examples are the parrotfish, angelfish, damselfish, Pomacanthus paru, Clinidae, and butterflyfish. At night, some change to a less vivid color. Besides colorful fish matching their environment, other fish (e.g., predatory and herbivorous fish such as Lampanyctodes hectoris, Holocentridae, and Pterapogon kauderni) as well as aquatic animals (Comatulida, Crinoidea, and Ophiuroidea) emerge and become active while others rest.

Other fish groups found on coral reefs include groupers, grunts, and wrasses. Over 4,000 species of fish inhabit coral reefs. Healthy reefs can produce up to 35 tons of fish per square kilometer each year, but damaged reefs produce much less.

Invertebrates

Sea urchins, Dotidae and sea slugs eat seaweed. Some species of sea urchins, such as Diadema antillarum, can play a pivotal part in preventing algae overrunning reefs. Nudibranchia and sea anemones eat sponges.

A number of invertebrates, collectively called cryptofauna, inhabit the coral skeletal substrate itself, either boring into the skeletons (through the process of bioerosion) or living in pre-existing voids and crevices. Those animals boring into the rock include sponges, bivalve mollusks, and sipunculans. Those settling on the reef include many other species, particularly crustaceans and polychaete worms.

Algae

Researchers have found evidence of algae dominance in locations of healthy coral reefs. In surveys done around largely uninhabited US Pacific islands, algae inhabit a large percentage of surveyed coral locations. The algae population consists of turf algae, coralline algae, and macroalgae.

Seabirds

Coral reef systems provide important habitats for seabird species, some endangered. For example, Midway Atoll in Hawaii supports nearly three million seabirds, including two-thirds (1.5 million) of the global population of Laysan Albatross, and one-third of the global population of black-footed albatross. Each seabird species has specific sites on the atoll where they nest. Altogether, 17 species of seabirds live on Midway. The short-tailed albatross is the rarest, with fewer than 2,200 surviving after excessive feather hunting in the late nineteenth century.

Sea snakes feed exclusively on fish and their eggs. Tropical birds such as herons, gannets, pelicans, and boobies feed on reef fish. Some land based reptiles intermittently associate with reefs, such as monitor lizards, the marine crocodile, and semi-aquatic snakes like Laticauda colubrina.

Schooling reef fish

Caribbean reef squid

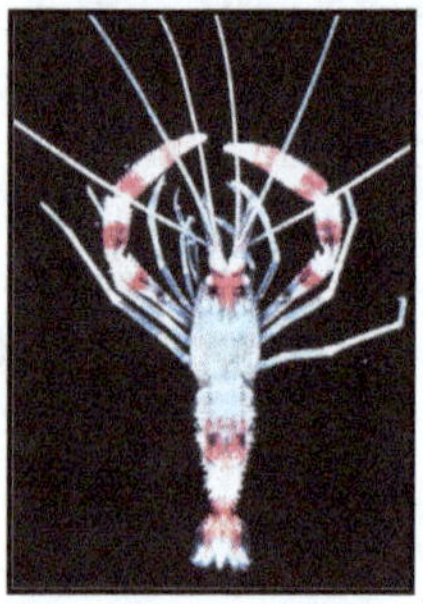

Banded coral shrimp

The whitetip reef shark almost exclusively inhabits coral reefs

Green turtle

Giant clam

Soft coral, cup coral, sponges and ascidians

Banded sea krait

Threats to Coral Reefs

A number of forces threaten the survival of coral reef organisms, as well as the structural integrity of the reefs themselves. Many coral reefs are plagued by predatory species, bleaching, and the effects of various human activities.

Crown-of-thorns Starfish

Certain biological factors, such as the fish and invertebrates that feed on the soft tissues of reef builders and the organisms that bore into coral rock, may contribute to the destruction of coral reefs. One of the most destructive creatures known is Acanthaster planci, the crown-of-thorns starfish, which during the 1960s multiplied spectacularly and removed the soft tissues from large areas of many reefs in the southwest Pacific. A. planci feeds by everting its stomach and liquifying and absorbing the tissues of the corals. By the late 1970s it had become apparent, however, that the sudden spread of A. planci was part of the organism's natural life cycle and that the coral reefs could regenerate rapidly after such an infestation. Coral-rock borers include boring algae, boring sponges (of great significance), various polychaete and sipunculid worms, and many bivalves and a few gastropods. These organisms usually penetrate the rock mechanically but in some cases do so chemically. Extensive damage is caused both by their own activities and by the assistance they give to the erosive action of the sea.

Crown-of-thorns starfish (Acanthaster planci)

Coral Bleaching

A phenomenon known as bleaching caused extensive devastation among coral reefs in the east Pacific since the early 1980s and in the Caribbean since the mid- to late 1980s. It is called bleaching because zooxanthellae (which are golden-brown in colour) are expelled, leaving the white coral exposed. While the cause of bleaching is not yet fully known and extensive research is under way, it is believed that the most likely factor is unusually high seawater temperatures, approximately 30 °C (86 °F). Other factors linked to coral bleaching include rising acidity associated with the increased absorption of carbon dioxide by the oceans, ultraviolet radiation exposure, drying from extreme low-tide events (such as those associated with El Niño), sedimentation, pollution, and disease.

Other Threats

A living coral reef may have economic potential as a major tourist attraction. By touching corals and disturbing the other animals that live there, visitors to the reef may interfere with the community dynamics of the reef system. Since coral reefs harbour many species of fishes and other sea life, local and commercial fishing can overharvest them. In their attempts to capture the animals of the reef, fishers may resort to using dragnets and explosives. Such practices scour or break up sections of the reef, destroying the corals and the numerous individual habitats they provide. Some coral reefs may be cloaked by excess sedimentation from terrestrial erosion. "Smothering," as this is called, may prevent reef plants from obtaining adequate sunlight or may promote the growth of harmful algal blooms.

Threats from Anthropogenic Factors

Island with fringing reef off Yap, Micronesia. Coral reefs are dying around the world

Coral reefs are dying around the world, threatened by a diversity of human activities. In particular, coral mining, coastal development, agricultural and urban runoff, pollution (organic and non-organic), overfishing, blast fishing, disease, and the digging of canals and access into islands and bays are localized threats to coral ecosystems. Broader threats are sea temperature rise, sea level rise, and pH changes from ocean acidification.

In El Nino-year 2010, preliminary reports show global coral bleaching reached its worst level since another El Nino year, 1998, when 16 percent of the world's reefs died as a result of excessive water temperature. In Indonesia's Aceh province, surveys showed some 80 percent of bleached corals died. In July, Malaysia closed several dive sites after virtually all the corals in some areas were damaged by bleaching.

General estimates show approximately 10% world's coral reefs are already dead. According to an assessment released in 2011 by a network of more than 25 organizations, including Nature Conservancy, the Global Coral Reef Monitoring Network, and UNEP-WCMC, an estimated 75% of the coral reefs of the world are threatened by anthropogenic impacts, and more than 90% will be threatened by 2030 if these threats are not dealt with in a timely fashion.

Mangrove Ecosystem

The term "mangrove" refers to a tidally influenced wetland ecosystem within the intertidal zone of tropical and subtropical latitudes. Mangrove also designates the marine tidal forest that includes trees, shrubs, palms, epiphytes and ferns. The distinctive community of plants and animals associated with mangroves is sometimes referred to as the 'mangal'. The origins of the terms for mangrove are well researched by Vanucci. She concludes that the word mangue is from West Africa, Senegal, Gambia and Guinea. The English word mangrove is a derivation of the Portuguese or Spanish meaning grove made of mangue. Mangrove ecosystems are heterogeneous habitats with an unusual variety of animals and plants adapted to the environmental conditions of highly saline, frequently inundated, soft-bottomed anaerobic mud. Not all mangroves are obligated to live in saline intertidal areas. Plants that are confined to the mangrove are called true mangroves; plants that can also occur elsewhere are called mangrove associates. Mangrove associates never grow in true mangrove communities and may occur in terrestrial vegetation. The mangrove fauna includes terrestrial, marine, temporary and permanent animal species, all of which have different adaptations to cope with the mangrove environment. The diversity of mangroves is high, but the variety of mangrove ecosystems also makes it difficult to produce general guidelines for conservation and management of mangroves because each system is unique.

Mangroves are found in depositional coastal environments where fine sediments, often with high organic content, collect in areas protected from high energy wave action. Mangroves are salt-tolerant plants of tropical and subtropical intertidal regions of the world. The specific regions where these plants occur are termed as 'mangrove ecosystem'. These are highly productive but extremely sensitive and fragile. Besides mangroves, the ecosystem also harbours other plant and animal species.

Mangrove forests are an ecosystem of incredible biological diversity comprising hundreds of algae, mollusk, crustacean, fish, insect, reptile, bird, and mammal species. Saltwater crocodiles, pelicans, egrets, spoonbills, flying foxes and monkeys are all residents of mangroves, just to name a few. The trees' intricate network of roots provides a shelter for many oceanic and freshwater species, and the forest is a food source for many types of organisms as well. While certain species spend their entire life span in this ecosystem, numerous fish species use the shelter of the mangrove as a nursing ground before heading to the open ocean, to the extent that one third of all marine species were born and raised in mangrove forests around the world. As a result, the mangroves' well being is central to the health of many more ecosystems.

Mangrove Species

Mangroves—forests of tropical trees and shrubs rooted in saltwater sediments between the coast and the sea—are crucial nurseries for coral reef fish, according to a new study John Roach, 2004). The tallest and most distinctive mangrove is the red mangrove (Rizophora mangle). Under optimal conditions this tree can reach 80 feet in height in some parts of the world, but on average reaches a height of 20 feet. The distinctive characteristic of this species is its reddish prop roots which can extend out of the water up to 3 feet. These roots are the primary home for the many species of organisms that thrive in the ecosystem. The red mangrove is the most tolerant of turbulent water. This mangrove is the first specie to meet the aggressive waves from the open ocean. The black mangrove (Avicennia germinans) is smaller and grows closer to the shoreline in areas slightly higher in elevation, more susceptible to tidal fluctuations. In the times of low tide pencil-like projections, or pneumatophores, extend into the atmosphere to provide oxygen for the underwater root systems. This specie can reach over 65 feet in some locations, but on average it will reach heights up to 50 feet. This tree appears similar to that of the red mangrove with opposite leaf arrangement. However the presence of pneumatophores and lack of prop roots sets them apart. The white mangrove (Laguncularia racemosa) is found at even higher elevations where soils are least saturated. The most distinguishing feature is the lack of aerial roots. Unlike the black and red mangroves the white mangrove conceals its roots underground, with the exception of prolonged flooded areas where the tree will develop a sort of peg root to obtain oxygen. The fourth specie of mangrove is associated with the upland transitional zone is restricted in its range by its extreme sensitivity to cold. The buttonwood mangrove (Conocarpus erectus) is a shrub-like tree and is least like the other mangroves but still plays a vital role in the mangrove ecosystem being home to terrestrial animals and avian nesters.

Terrestrial Inhabitants

There is a wide range of mammals that visit the mangroves to hunt and forage for food. Some small like raccoons, opossums or agoutis, but there are also large animals that rely on these ecosystems. The very rare Florida panther was last seen in the everglades mangrove system, and due to habitat destruction it is estimated that roughly 50 individuals remain in the wild. The endemic key deer is only among the mangrove systems in the Florida Keys. In many other parts of the world monkeys are very common in mangroves, and in fact have been captured on video diving in the shallows collecting food. The greater flamingo and other special birds also visits these ecosystems were they forage in the shallows for tiny crustaceans. Many waterfowl species in North America spend their wintering months in and around the mangroves searching for food before their counter migration back north. To one particular sort of bird, mangrove canopies are among one of the few options for nesting. Pelagic birds, or sea birds, are birds that spend the majority of their lives at sea. These birds are highly efficient at staying aloft and use little energy when flying. They come to land rarely and often only to nest. To these birds the thick tangled canopy of the mangroves is ideal to build for nesting and rearing young. The seeds from a parent tree drift across the sea and rest upon an exposed section of sand, taking root and virtually creating a bird paradise. Black mangrove grows in the intertidal zone throughout the Gulf of Mexico. It will establish in nature from seed that floats and can travel some distance on the tides. These specialized birds, such as the magnificent frigate bird, can nest on an island free from predation and have an open buffet on either side to feed their young.

Aquatic Inhabitants

Among the tangle of the mangrove prop roots (Rizophora mangle) lays an absolute flourish of life. Nearby are beds of seagrass which are considered part of the mangrove ecosystem. The organisms from the mangroves interact with those of the seagrass beds. In the Caribbean, juvenile reef fish occupy the submerged prop roots of Rhizophora mangle and make frequent foraging runs into adjacent seagrass beds. Mangroves serve either as nurseries for juveniles or as feeding areas for transient fish and crustaceans. For many fish they start their lives in the adjacent sea grass beds, they migrate into the mangroves as they grow slightly larger, and then when they are just about fully mature they migrate across the flats to the coral reefs. This is known as ontogenic migration.

It has been found that certain coral reefs are correlated to certain mangrove systems. Therefore if a mangrove forest is removed, the fish populations of the correlating coral reef are negatively affected. Just as waterfowl seem to have a sixth sense when it comes to direction, reef fish also seem to know where their desired destination is as the set off on their final migration to the coral reef where they will live out the rest of their lives. Migrating fishes can move directly between mangroves and reef habitats, passing over deep water and seagrass beds. The reasons for undertaking such a stage-structured lifecycle are varied and include: (i) requirements for different food sources as the organism grows, (ii) changing risks of predation with size, such that sheltered habitats, where predator foraging efficiency is low, are chosen when the organism occupies its smaller, more vulnerable stages, and (iii) a need to reproduce in habitats which offer the greatest dispersal or survival of larvae. The maximum distance covered during migration, observed in a study by Peter J. Mumby in the cays of Belize, was 10 km.

Mangroves as a Nursery

There is a wide range of fish species that live in the nursery of the mangroves. There are some fish species that live in the mangroves fulltime, some come and go with the seasons, and some just live in the mangroves until they reach maturity. Species that live here full time are a variety of gobies, spotted trout, pipefish, and others. These full time residents never grow too large for their habitat and therefore never have reason to leave. During the rainy season, the increased flow of freshwater results in the appearance of freshwater species, such as the predatory Florida gar. This increased flush of freshwater has been known to drive some species further off shore in search of higher saline waters. Among the resident species and seasonal guests are the juvenile fish species which migrate to the coral reefs once they have reached maturity. Of these juvenile reef fish are various species of grouper, snapper, jacks, schoolmasters, and others. Among the grouper species that live in these habitats is the endangered Goliath grouper. These spectacular fish can reach weights of over 800 lbs. when mature. Goliath grouper breed in mass swarms in the open ocean and their fry grow in the seagrass beds and then they migrate to the nearby tangle of the mangroves. Goliath grouper remained in mangrove habitats for 5–6yrs (validated ages from dorsal spine sections), then emigrated from mangroves at about 1.0 m total length.

In this mangrove nursery there are many species of fishes that flourish in a predator reduced environment. The larger predatory fish are too large to pursue their prey in the mangroves. While within the mangroves themselves there are very few predators, in the more open waters are predator fish that await any small wanderers that may stray away from the protection of the mangroves. Fish like the tarpon, snook, bone fish, red fish, rays, and lemon shark work with the tides. They are fairly large fish and need deeper water to enter the shallows surrounding the mangroves. Once these predators get into the shallows they are on a race against the clock to hunt as much as they can before the tides retreat and they become stranded. These fish are opportunistic feeders taking whatever they can fit in their mouths; just about everything is fair game. These fish don't have any natural predators in these waters. However they are highly sought after by the saltwater angler.

Other Mangrove Inhabitants

The ecosystem of the mangroves is not just home to fish species. Also calling home to these parts is a wide variety of invertebrates. The extensive root systems, muddy bottoms, and open waters are all home to invertebrates that are well adapted to the temperature and salinity fluctuations as well as the tidal influences common to mangroves. These wide array of organisms feed on leaf litter, detritus, plankton, algae, and other small animals. Barnacles, mollusks, shrimp, crabs, lobsters, jellyfish, tunicates, etc. can all be found among the roots of the mangrove. There is also a very ancient animal that can be found living in the mangroves, the earliest fossils of the horseshoe crab dates back to roughly 450 million years ago.

Horseshoe crabs are scavengers and may be found among mangroves feeding on algae, invertebrates, and dead organisms. They are especially adapted to low oxygen waters, possessing up to 200 book gills used for respiration. On the edges of the mangroves and out in the seagrass beds one can find colorful anemones with their feeding tentacles sent up to catch food, sea urchins crawling about munching algae, and large gastropods known as conch scooting about the bottom. These large slow moving snails are known for their iconic shells, and have been collected for years for their meat. Also commonly found in these waters and collected for their meat is the Caribbean spiny lobster. Their habitat usage is similar to that of the juvenile reef fish. Depending on their maturity they sought different habitat type. Largely influencing their preference for habitat was predation. Predation on newly settled juveniles was greater in seagrass and coral crevices than in mangrove prop roots, whereas the survival of larger juveniles was higher in mangroves and coral patch reefs than in seagrass. These results suggest that mangrove habitats may function as a nursery for juvenile spiny lobsters but that the use of this habitat depends on shelter characteristics and the isolation of islands.

Along with invertebrates and fishes, sea turtles are also associated with mangrove vegetation, at least point in their life. The Atlantic ridley sea turtle is commonly observed in the mangroves that line the bays of south Florida. The loggerhead and green sea turtles have been known to utilize the mangroves as juvenile nurseries as well, receiving protection from predation and ample access to a food rich environment. The green sea turtle and the Hawksbill sea turtles have been observed feeding on the roots of mangroves as well as other submerged vegetation. Another reptile known to the mangroves are saltwater crocodiles. Mangroves are particularly well suited for juveniles. All around the world species of water snakes can be found living in mangroves, as well as some saltwater tolerant amphibians. The most common of these amphibians is the giant toad (bufo marinus). The giant toad, or more commonly known as the cane toad, is native to Central and South America but has been introduced in the Caribbean and some Pacific Islands. Originally introduced to control the cane beetle it is now considered a pest. There are also two quite different aquatic animals that also are very common in and around mangroves. Feeding on the small fish species dolphin are quite common to mangroves. Feeding on seagrass and submerged vegetation manatees are also often seen swimming around the edges of mangroves along the coast of Florida. Another species of manatee that can be seen in coastal marine habitats is found in West Africa.

Erosion Resistance

Besides for crucial for wildlife habitat both above and below the surface, mangroves also serve a purpose to the land beyond them. Mangroves are very efficient at counteracting erosion. In many areas around the world mangroves are removed for various reasons, and in these areas the effect of erosion can be seen almost immediately. Mangroves act as a natural wave block from the strong waves that have built up on the open ocean. Even a small band of mangrove is an efficient blocking waves. Mangroves are extremely important in times of increased wave strength, such as during tropical storms. The best example on finds is the super-cyclone which occurred on the 29th October 1999 with a wind speed of 310 km hr-1 along the Orissa coast (India) and played havoc largely in the areas devoid of mangroves. On the contrary, practically no damage occurred in regions with luxuriant mangrove growth (K Kathiresan). This example along should be evidence enough to prove that mangroves are essential in protecting coastal lands. Ironically the highly populated areas most in need of protection from tropical storms are the area's most likely to remove mangroves. A prime example of this would be Cancun, Mexico. Cancun is situated at the very tip of the Yucatan Peninsula, which is a relatively large piece of land situated in the middle of the Caribbean Sea. Given its location, the peninsula is highly susceptible to hurricane activity and with the removal of their mangroves they are rendered helpless to any strong tropical storm. The logic for the removal is for more pleasing aesthetics, yet less aesthetically pleasing are the sandbags set in place to combat the waves and high water. Without the mangroves to hold sediment in place, even the gentlest of waves will take some amount of sediment out into the ocean. A wellknown and credible institute has witnessed this first hand. The Smithsonian Institute operates a field station on a small island, 0.4 hectare in size, 14 miles off the cost of Dangriga, Belize. The Carrie Boy Field Station is located on the Meso-American Barrier Reef, and is devoid of mangroves. Do to this absence of mangroves the already small island shrinks every year, at an average decreasing rate of 2 inches. Along with keeping sediment in place, mangroves also filter the water passing through the sediment. "Black mangrove is valuable in restoring brackish and salt water marshes due to its ability to filter and trap sediments. Mangrove forests, which include black mangrove, have a high capacity as a sink for excess nutrients and pollutants. It also mixes well with other native plants to reduce wave energy". Mangroves are so efficient at filtering nutrients from water that in some places they are being used on a municipal level to treat sewage. A mangrove wetland was constructed in Futian, Shenzhen, China to do just this. Three belts of mangroves were planted, housing three different species. The belts through which the water was fed held dimensions of 33 m in length, 3m in width, and 0.5 m in depth. Expectedly the removal of organic matter and nutrients were positively correlated with plant growth. The results indicated that mangroves could be used in a constructed wetland for municipal sewage treatment, providing post-treatment to remove coliforms was also included. In the natural setting filtration is key as well, especially in areas where development is prevalent. The filtration of runoff is particularly important where the surrounding region contains sea grass or coral reefs. These habitats are particularly vulnerable to deterioration if water quality declines. If coral reefs and sea grass habitats were to be lost, numerous highly valuable ecosystem goods and services would also be lost.

Economic Importance

Mangroves are economically important as well. The value of mangrove ecosystem services worldwide has been estimated as an annual global flow of US$ 1,648 billion. The problem is when harvest exceeds production. Of 2500 households in 21 rural communities (about 13.000 people) near the Caeté estuary, 83% derive subsistence income, and 68% cash income through use of mangrove resources. Near this estuary on the northern Brazilian coast, everything is used in the ecosystem from the sea, to the land, and even the wood itself. The mangrove crab (Ucides cordatus) is collected and sold by 42% of households, and constitutes a main income source for 38%. Including processing and trading occupations, over half of the investigated populations depend on the mangrove crab for financial income. In addition to the mangrove crab in these regions, many households depend on the use and sale of the mangrove wood and bark. Not only in Brazil but virtually worldwide mangrove wood is collected on both the small scale and commercial scale. Mangrove wood is high in tannin therefore is very sought after for its durability in the timber industry. As with any wood harvest it is done both sustainably and not sustainably. In fact, mangroves can be very successfully managed for sustainable timber production – the Matang Mangroves in western Peninsular Malaysia are a classic example. The management system, first started at the beginning of the 20th century, is based on a 30 year rotation. Small areas are clear-felled in a patchwork manner, and allowed to regrow before the next harvest. Mangrove wood burns with a high heat output and no smoke, making it ideal fuel wood. Coastal rural inhabitants have been using mangrove as fuel for many generations. Consequently these characteristics make it very attractive for mass collection coal production. One ton of mangrove wood is equivalent to 5 tons of Indian coal. Due to its durability the rayon industry also ravages the mangroves for their wood, as does the wood chip industry. These three industries are highly responsible for the destruction of mangroves. The pneumatophores can be used to make bottle stoppers and floats for fishing. A perhaps unexpected industry is the mangrove honey industry. During full bloom bees and other insects are found buzzing about mangroves as they are any patch of wild flowers. And in doing so collect pollen on their bodies which is later brought to the hive and put to use. Many people in foreign lands know of the sweet nectar produced and harvest this honey produced. For instance, the Sundarbans provide employment to 2000 people engaged in extracting 111 tons of honey annually and this accounts for about 90% of honey production among the mangroves of India. In Bangladesh, an estimated 185 tons of honey and 44.4 tons of wax are harvested each year in the western part of the mangrove forest.

Fisheries

Perhaps the most exploited, but economically profitable, parts of the mangrove ecosystem are the fisheries. On a subsistence level, fishing can be done sustainably and is quite crucial for some populations of people. On a local level many individuals dive for the Conch and keep themselves as well as sell them to local restaurants. But on a commercial level, animals are taken out of the mangroves for money and often at a rate at which the ecosystem cannot supply for the demand. Most common are shrimp and the various species of fish, although other crustaceans and shellfish are also sought after. Implementing the proper techniques the fisheries can be harvested from on a sustainable level. Most of the time the fishermen are often overwhelmed by greed, which ultimately leads to the down fall for the fishery. And when the ecosystem does not provide enough for what the demand is, aquaculture is established. Typically this means that the natural mangrove ecosystem is destroyed, and what remains is manipulated to better suit for a certain organism in order to gain cash profit. Consequently this only suits those particular organisms banishing the others from existing there. The food and other chemicals put into the water also adversely affect the rest of the surrounding waters around these man made ponds. The driving force is the high demand for shrimps in the developed countries and the governments of developing countries grossly undervaluing their mangroves.

Carbon Sequestration

Beyond the economical possibilities and the crucial wildlife habitat for so many varieties of organisms, is yet an even more important value of the mangrove. This is one that is important not only in small communities or even whole countries, but on a global scale. This unique characteristic is the mangroves ability to remove carbon from the atmosphere. Much like the wetlands of the inlands and the boreal forests of the temperate climates, mangrove systems are referred to as carbon sinks. The ocean itself does a great job through its direct air to sea exchange activity. This more or less is the action of waves capturing air and bringing the carbon into the ocean. With the combined oceanic air-to-sea exchange and the mangroves ability to seize atmospheric carbon, the two seize a huge amount of carbon. Measurements show that mangroves have the ability to seize around one and half tons of carbon per hectare per year. This is approximately equivalent to the amount of carbon a motor vehicle releases to the atmosphere each year (assuming each car uses approximately 2,500 liters of petrol per year). Indonesia has 4.5 million hectares of mangrove, and from this previous information we can conclude that the emission from 5 million cars can be erased by the mangrove ecosystem in Indonesian alone. A high amount of carbon is stored in the upper layer of mangrove sediment, around 10 %. Each hectare of mangrove sediment consequently contains about 700 tons of carbon per every meter of depth, and digging up the sediment to create ponds for aquaculture results in the potential oxidation of 1/400 tons per hectare per year. Naturally only about half of this would become oxidized over the span of about ten years. The sequestration rate is then catalyzed to a rate of around 50 times the natural rate, thus eliminating the mangroves ability to absorb atmospheric carbon.

Kelp Forest Ecosystem

Kelp is the common name for any of the large seaweeds comprising the order Laminariales within the brown algae (class Phaeophyceae). There are about 30 different genera of kelp, largely found in colder seas along coastlines in the Northern Hemisphere. Some species grow very long and form kelp forests. Kelp forests are underwater areas in temperate and polar coastal oceans with a high density of kelp and are recognized as one of the most productive and dynamic ecosystems on earth.

Ecologically, kelp provide a physical substrate and habitat for animals, as well as food for such aquatic organisms as sea urchins, seastars, isopods, kelp crabs, and herbivorous fishes. For humans, kelp has long been used as a source of nutrition, providing such elements as iodine, calcium, iron, and diverse vitamins, and its use in medicine traces as far back as 3,000 B.C.E. in China. Kelp often is recommended today by naturopaths and herbalists for various conditions, such as thyroid disorders. Kelp is particularly important as part of present-day Japanese cuisine.

Through the nineteenth century, the word "kelp" was closely associated with seaweeds that could be burned to obtain soda ash (primarily sodium carbonate). The seaweeds used included species from both the orders Laminariales and Fucales. The word "kelp" was also used directly to refer to these processed ashes.

Kelp is a type of seaweed, often large, within the order Laminariales of the brown algae. Seaweeds are macroscopic, multicellular, marine algae, which generally are benthic, being anchored to the bottom of the ocean or to solid structures. Brown algae comprise a large group of multicellular, mostly marine algae. Brown algae are placed in the class Phaeophyceae, but their higher level classification is not settled, being variously placed in in the kingdoms Plantae, Protista, Chromista, or Chromalveolata, and in the Divisions Heterokontophyta or Phaeophyta.

Giant kelp in the Monterey Bay Aquarium

Kelp does have a plant-like appearance, having tentacle-like roots from which grows a slender stalk with long, flat, leaf-like blades. The body of an individual organism is known as a thallus rather than as a plant (Kingdom: Plantae). The morphological structure of a kelp thallus is defined by three basic structural units:

- The holdfast is a root-like mass that anchors the thallus to the sea floor, though unlike true roots it is not responsible for absorbing and delivering nutrients to the rest of the thallus.
- The stipe is analogous to a plant stalk, extending vertically from the holdfast and providing a support framework for other morphological features.
- The fronds are leaf- or blade-like attachments extending from the stipe, sometimes along its full length, and are the sites of nutrient uptake and photosynthetic activity.

In addition, many kelp species have pneumatocysts, or gas-filled bladders, usually located at the base of fronds near the stipe. These structures provide the necessary buoyancy for kelp to maintain an upright position in the water column.

Growth occurs at the base of the meristem, where the blades and stipe meet. Growth may be limited by grazing. Sea urchins, for example, can reduce entire areas to urchin barrens. The kelp life cycle involves a diploid sporophyte and haploid gametophyte stage. The haploid phase begins when the mature organism releases many spores, which then germinate to become male or female gametophytes. Sexual reproduction then results in the beginning of the diploid sporophyte stage which will develop into a mature plant.

Characteristics of Kelp Forests

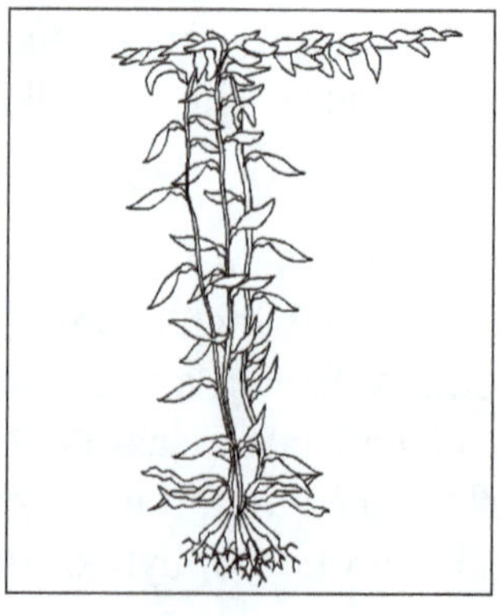

Kelp forests have some characteristics in common with forests found on land. They are complex ecosystems that support many interconnected food webs. They provide a variety of habitats for animals at all trophic levels. They can be divided into layers including the canopy, understory, and forest floor. They progress through seasonal cycles.

The forest layers help describe this complex ecosystem. The uppermost layer of the kelp forest is called the canopy. Along the California coast, the giant kelp is the main algae found in this layer. Among the fronds in the kelp canopy live bryozoans, hydroids, mysids, amphipods, isopods, crabs, snails, nudibranchs, and juvenile fish of all sorts. Some of these animals feed directly on the kelp and are capable of considerable damage if not kept in control. Controlling them are the predatory fish usually found in the mid-level of the forest. Kelp surfperch, pipefish, clingfish, tubesnouts, and giant kelpfish are active predators here. Kelp bass and various rockfish often prey on the smaller fish. Kelp crabs and kelp snails feed among blades in the canopy and often fall prey to seabirds. The density of kelp fronds in the canopy affects the amount of light reaching the bottom and so determines the types and abundance of algae on the reef.

The understory lies below the canopy. This layer has algae of medium height much like the bushes and shrubs of the land forest. It also supports a variety of food webs. These food webs can be found in open water pathways, on reef tops, or in cracks and crevices. Shorter kelp varieties such as oarweed and featherboa are characteristic of this layer. The forest floor is the bottom layer. It includes the substrate and benthic creatures. In the kelp forest, this layer contains many animals. Here tunicates, sponges, and anemones are abundant.

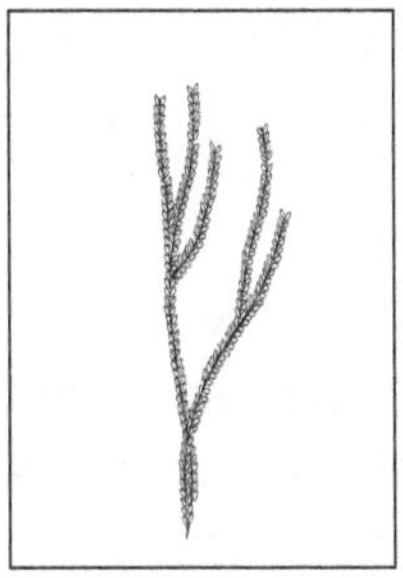

The kelp forests progress through seasons as the amounts of sunlight and nutrients change with natural cycles. The spring season begins as March winds blow surface waters offshore. This creates upwelling as nutrient-rich waters are brought up from the depths. At the same time, days are lengthening. More sunlight increases kelp growth. The kelps continue to grow rapidly until summer. In the top layers, long days and cool nutrient-rich water create the perfect environment for phytoplankton. These ocean producers become abundant and feed many animals living in this area. As the canopy thickens, light is cut off to the lower layers. Growth at these levels begins to slow. As fall approaches, wind directions change. This stops the upwelling of nutrients. The lack of nutrients and sunlight slow the growth of the forest. The kelp plants become weakened and begin to thin. The dead blades or kelp shed are torn apart by winter storms. Animals growing on the kelp drift are swept away with the dying fronds. Holdfasts also begin to weaken and loosen. They roll about driven by the moving water. The forest becomes sparse as winter progresses. This allows the sunlight to again penetrate to lower levels. As the months pass, days grow longer and winds become favorable. A new spring season begins.

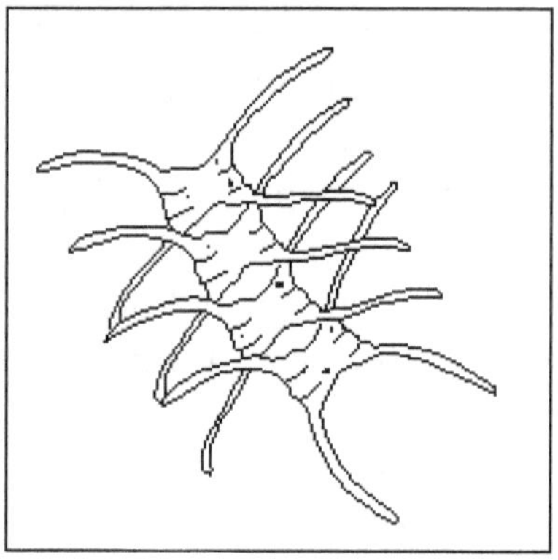

Distribution

Kelp forests are mostly found in cool, shallow, nutrient-rich water near coasts. Kelp is a brown alga (Phaeophyceae), which requires access to light in order to photosynthesize - this is the reason for their abundance in shallow coastal waters. Kelp is one of the fastest growing organisms in the world, growing anywhere from 2 centimeter to 1 meter each day, depending on species and environmental conditions.

Kelp can be found on the coast of every continent besides Antarctica Kelp forest ranges include areas of cold water, generally coinciding with coastal upwelling. The eastern Pacific Ocean along the west coast of North America, is an ideal area for kelp forests due to the presence of upwelling bringing cold water and nutrients to the surface from deep ocean water. Towards the poles, kelp forest development is limited by the availability of light at higher latitudes, while kelp towards the equator tends to be limited by the presence of warmer waters.

Ecology Habitat

Kelp forests are an important foundation species, which provide large areas of habitat within the ocean. Kelp forests have a wide variety of inhabitants within their boundaries: invertebrates frequently graze on the blades fish find shelter in the fronds, and invertebrates - such as brittle stars, sea stars, anemones, sponges and tunicates - live in the holdfast. Kelp is also the single largest source of fixed carbon within their ecosystems. Pieces of decomposing kelp sink deeper into the ocean, providing nutrients to deep sea organisms. Thus, kelp derived particulate organic matter (POM) is an important food source for fauna both within and outside the kelp forest.

Since kelp is capable of attenuating waves and dampening their energy, many animals, such as mammals and birds, are able to seek refuge in the forests. Various species of kelp tolerate ocean storms differently. In fact, along the California coast giant kelp has been known to dominate in years with less turbulent sea conditions, while bull kelp is more dominant in open waters and years of turbulent conditions.

Sea Otters and Urchin Barrens

Sea otter (Enhydra lutris) are one of the keystone species within kelp forests - meaning that they have a substantial impact within an ecosystem despite their relatively small population. Otters' diets consist mainly of invertebrates such as purple sea urchins (Strongylocentrotus purpuratus) which inhabit kelp forests along North America's west coast. Urchins feed on kelp and can potentially damage kelp forests by chewing through kelp holdfasts. Otters are the primary predator of urchins, keeping their populations at sustainable levels and maintaining the ecological balance of the kelp forest. Without otters a trophic cascade can occur; leading to destructive levels of grazing by urchin, causing a phenomenon known as an "urchin barren". Due to kelp's role as a foundation species, significant populations of organisms that depend on their habitat are negatively impacted.

Rocky Shore Ecosystem

Rocky shores are found where the sea meets the land. They support a diverse mix of plants and animals which have adapted to survive this habitat's unique conditions. Along the exposed coast of Queensland, constant wave action and the rise and fall of tides can make these shores tough places to live. Waves are less of an issue to animals and plants of rocky shores in other areas such as estuaries, the sheltered side of the great sand islands and within the Great Barrier Reef lagoon.

As well as supporting lots of unusual plants and animals, rocky shores are important fish nurseries and roosting and feeding grounds for birds. Along with their commonly associated algal beds, they also help stabilise inshore sediments.

Because big parts of rocky shores are exposed at low tide, they're great places to study marine life, though you need to take care of yourself (especially on high wave action coastlines) as well as taking care to minimise your impacts on this sensitive environment.

Rocky shore ecosystems are coastal shores made from solid rock. They are a tough habitat to live on yet they are home for a number of different animals and algae. Rocky shore ecosystems are governed by the tidal movement of water. The tides create a gradient of environmental conditions moving from a terrestrial (land) to a marine ecosystem.

Ecosystems on rocky shores have bands of different species across the intertidal zone. The distribution of different species across the rocky shore is influenced by biotic and abiotic factors from above high tide to the sub-tidal zone. Different species are adapted to different environmental conditions. Some organisms can withstand being exposed to the sun for most of the day and live in the upper parts of the rocky shore. Other organisms need to be covered by the tide for most of the day and are only found lower on the rocky shore.

The different sections of the rocky shore are exposed to varying amounts of stress. Stress may be caused by an increased time out of water and in the Sun, more pressure from predation or grazing, or other factors that change across the intertidal zone. As the conditions are changing over the rocky shore ecosystem, various bands of different organisms are found along the intertidal zone.

Above the Intertidal Zone

The littoral/intertidal zone can be divided into different zones across the rocky shore that each have similar environments. Above the intertidal zone sits the supralittoral zone which is virtually a terrestrial environment. Species found within the supralittoral zone must be tolerant to high levels of salt caused from wave spray. Here it is common to find periwinkles (little snails with elongated shells) and lichens on the rocks but the diversity of species in the supralittoral zone is fairly limited.

Intertidal Fringe

Between both the supralittoral and intertidal zones is the littoral fringe or the intertidal fringe. This is the area where the high tide may or may not be covering the rocks depending on the height of the tide. Tide heights vary depending on the placement of the Moon, the Sun and atmospheric pressure. The diversity and abundance of animals and plants are still low in the littoral fringe and periwinkles are still the animals you are most likely to find.

Intertidal Zone

Below the littoral fringe is the most diverse and interesting area of the rocky shore, the intertidal zone. The regular covering and uncovering of the shore by the waves provides a regular income of food and nutrients for plants and animals. The density of plant and animal communities in the intertidal zone are often very high. Due to the high levels of nutrients, the distribution of many species is limited by competition with other species for space.

Rocky Shore Invertebrates

The animal populations on the rocky shore are dominated by invertebrates such as barnacles, mussels, oysters, tubeworms, limpets, chitons, snails, crabs and starfish. Barnacles, mussels, and oysters are all stationary filter-feeders. They obtain their food by filtering for phytoplankton and edible particles straight from the water above them while they are submerged beneath the tide. They are often the dominant animal species on the rocky shore but their abundance is influenced by competition with each other, the environment of the shore, and predation pressure from predators like starfish and oyster borers.

Limpets, chitons, and various gastropods (snails) belong to another very important group of animals that live on the rocky shore – the mobile grazers. These grazers move around the intertidal zone and scrape the rocky shore free of any algae and settled juvenile shellfish. By scraping off algae they maintain the structure and diversity of rocky shore ecosystems by preventing algae from dominating the intertidal zone.

The large abundance of food on the rocky shore attracts a number of predator's species that prey on the slow-moving grazers and stationary filter-feeders. Starfish and gastropods are the most effective predators on the rocky shore and have significant abilities to control grazer and filter-feeder abundance.

Their control over these populations makes these predators important regulators of the structure of the rocky shore ecosystem. Feeding when the tide is in, these rocky shore predators are able to break through the hard exterior shell of many slow many species that are common on rocky shores.

Other predators include crabs which are the fastest moving animals on the rocky shore. Their superior mobility comes at a sacrifice. Crabs have a reduced ability to stick to the rocky substrate and subsequently must behave in a way to reduce the risk of being knocked around by incoming waves.

Algae

It is also common to find a number of algae on rocky shores and your chances increase as you move lower down the rocky shore. Seaweeds are most abundant in the sub-tidal zone where they often form dense kelp forests and provide habitats for many fish, worms, crustaceans, gastropods, and many more marine animals. Certain species can be found in the intertidal zone and more commonly these are green species of algae such as sea lettuce.

Birds and Mammals

The rocky shore is also visited by many birds and marine mammals as a place to rest, warm up and to breed. Many seabird species, such as penguins, shags, gannets, and albatross, use the rocky shore as a nesting area and to dry off and warm up after a cold swim. If you are lucky enough to live or visit an area with seals or sea lions, chances are you will find them basking in the sun on the rocky shore, warming up after a cold swim.

Factors Influencing Rocky Shore Habitats, Plants and Animals

Of the many factors that influence habitats, plants and animals on the intertidal rocky shores, energy forces (mainly as wave energy) and tidal inundation are very significant. Another factor or attribute influencing rocky shores includes the composition of the rock, which can determine how the rock breaks up into smaller components (e.g. boulders, cobbles, pebbles, gravel etc.)

Waves

Wind blowing across the ocean forms peaks and troughs in the water surface which appear to travel as a wave. Along the exposed southern coast of Queensland, waves can travel a long distance, and accumulate enough energy to be quite large. Waves break over rocky shores and plants and animals living on these places have adapted to being pounded by waves. Where waves splash higher on the rocky shore, animals take advantage of this splash zone.

In bays and estuaries, in the sheltered side of large islands (e.g. Fraser, Moreton and Stradbroke Island), and within the sheltered waters of the Great Barrier Reef lagoon, waves are generally smaller and more gentle. There is less wave splash and the intertidal area is mainly influenced by tidal inundation. Waves can also be created by large storm events, such as cyclones and east coast lows which are exceptions to the rule for high energy and low energy rocky shores.

Tides

The tide's rise and fall is one of the main factors affecting life on rocky shores. When the tide falls, plants and animals on rocks are exposed to air. They must develop special adaptations to survive until the tide comes in again. When high tides aren't very big, plants and animals which live high on the shore may be exposed to air for several days. Organisms which live very low on the shore may only occasionally be exposed to air.

Tides can also act as an energy force over the tidal cycle, such as during tidal ebb or flow when the water is rushing away from or towards the shoreline.

Nature's Survival Tricks

Special adaptations enable animals and plants to live in these conditions. Many animals avoid sun, drying air and predators such as birds, by staying in cracks, under rocks or in their own burrows at low tide. Some sessile animals such as barnacles and oysters close their valves tightly to avoid drying when the tide goes out. They come out to feed when covered by water.

Others may be well camouflaged and appear to be invisible. Organisms which are constantly pounded by waves are often very tough, so it doesn't matter if they're hit hard. Others are flexible or flat, so they bend instead of breaking when they're hit, or don't get hit at all.

Rocky Shore Dwellers

Seaweeds

Most plants found on rocky shores are seaweeds. They're algae which mean they can live on hard surfaces where plants with roots wouldn't be able to survive. Instead of roots, they have special suckers called 'holdfasts' which cling to rock, even in big waves. They don't have flowers, or normal stems or leaves. The bit that looks like a stem on some seaweed is called a 'stipe'. This joins the holdfast to the 'frond', the leaf-like part. Seaweeds are mainly green, red or brown, depending on which wavelength of sunlight they're trying to trap. Not all seaweeds have long, floaty fronds. The fronds can be tiny, so the seaweed looks like velvet covering the rocks. Other may look like tiny cabbage leaves. The kinds of seaweeds that grow on high energy rocky shores of southern Queensland may differ from those of the low energy shores of bays and estuaries and further north within the Great Barrier Reef lagoon.

Lichens

These black, orange, yellow or grey plants are actually made up of a fungus and a microscopic algae living together and sharing food and energy to grow. They live at the top of the shore where the tide doesn't rise and are the only plants on a rocky shore which are not algae.

Microscopic Plants and Cyanobacteria

Some rocks along the shore look bare. But that doesn't mean there are no plants living on them. If you look carefully at rocks in mid- to high tidal level, you'll notice they're often yellowish or pinkish. This is because they're covered with microscopic plants, many of which are diatoms, tiny, single-celled plants with hard silica shells. These plants are the main food for many grazing animals on rocky shores. Cyanobacteria, previously known as blue-green algae, can also grow, showing up as dark shadowy colours on the rocks.

Grazing Snails, Limpets and other Molluscs

Many species of these animals live on rocky shores. They eat microscopic plants, lichen or seaweed, depending on which part of the shore they inhabit. Limpets are snails which have a cup-shaped shell instead of a coiled one. They use a large, flat foot to tightly clamp the rock. Snails and limpets of high energy shores may differ from those of low energy shores where they do not need to hang on so tightly. On low energy shores, oysters or mussels can form large reefs of their own in the intertidal and subtidal areas, full of nooks and crannies for animals to live.

Barnacles

Barnacles attach themselves to one spot on the rocky shore and never move, not even to feed. Using specialised legs, they catch food as it floats by in the waves. Barnacles' shells are made of several plates. Their size, shape and position on the shore depends on whether they're exposed to big waves and whether they are splashed by spray from the waves. Barnacles can be present in large numbers, occupying distinct zones on high energy rocky shores.

Sea Squirts

Also known as cunjevoi, sea squirts are similar to barnacles in that they're both filter feeders which stay in one spot. Cunjevoi are usually found on high energy rocky shores. These animals pump large amounts of water through their bodies while under water, and then filter the food out. Since they store a lot of water during low tide, they squirt when you step on them.

Sea squirts are an important food source and habitat, so if you destroy cunjevoi to use them as bait for fishing, you're also wrecking the lives of many other creatures.

Anemones

The scientific name for anemones and corals is Anthozoa - Greek for 'flower animals'. Although they are animals, anemones grow in forms which resemble plants. An anemone is a single animal with a sack-like body and tentacles around a single opening. Some are capable of sliding about very slowly. In rock pools and on reefs just off shore, there are many species of anemone.

Corals

Hard and soft corals can occur in rock pools or in the lowest tidal areas in the warmer waters of the Great Barrier Reef lagoon as well as on the sheltered lee side of exposed rocky shores and platforms. These may include subtropical species that cover parts of the rocks, taking the shape of the rock; or they may be mainly tropical corals that form significant limestone structures of their own. Some hard corals look obviously stony, yet others have soft polyps which come out in the day, hiding their stony skeleton. Soft corals also come in similar leathery or feathery forms which hide the basic polyp structure.

Importance of Rocky Shores

There are many places for animals to live among the rocks, platforms, depressions, cobbles, pebbles and boulders. As well as providing homes for many animals, rocky shores are a productive food source and an important nursery area for many fish and crustacean species. This habitat also provides lots of food for fish. The commercially important fish found around rocky shores include blackfish, yellowfin bream, snapper, tarwhine, trevally, yellowtail and sampson fish.

Some of these species like to shelter by rocky shores, in areas where stands of seaweeds break the waves' power. Algal beds of this habitat are an important food source for rare and threatened species like marine turtles. And at low tide, wading birds love to feed on crabs and limpets on exposed rocks.

Sandy Beach Ecosystem

Sandy beach in Middelkerke - Belgium

Sandy shores or beaches are loose deposits of sand, including some gravel or shells, that cover the shoreline in many places. They make up a large portion of the world's ice-free coastlines. Beaches serve as buffer zones or shock absorbers that protect the coastline, sea cliffs or dunes from direct wave attack. It is an extremely dynamic environment where sand, water and air are always in motion. Beaches also provide important coastal recreational areas for a many people. Fine-grained sand beaches tend to be gently sloping and quite flat.

Formation

Sandy beaches are soft shores formed by deposition of particles that have been carried by currents and waves. The transported material is in part derived from shore erosion, but the major part is generally derived from land and transported by rivers to the sea in the Holocene era or earlier. A more comprehensive introduction to beach formation is given in Coastal Hydrodynamics And Transport Processes. The two main types of beach material are quartz (=silica) sands of terrestrial origin and carbonate sands of marine origin. The carbonate sand is weathered from mollusk shells and skeletons of other animals. Other material includes heavy minerals, basalt (=volcanic origin) and feldspar.

Characteristics

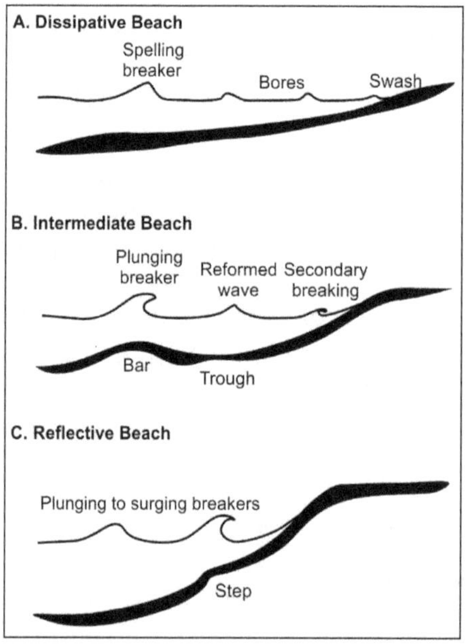

Different beach types based on morphodynamic scale

The grain size of sand varies from very fine to very coarse. The particle diameter is shown in the table below. Quartz sands have a slightly lower density (~2.6g.cm^{-3}) than carbonate sands (2.7–2.95g.cm^{-3}). The quartz particles are generally more rounded. Calcium carbonate particles sink more slowly in water due to their more irregular shapes, even if their density is higher.

Generic Name	Particle Diameter (mm)
Very coarse	1.0 to 2.0
Coarse	0.50 to 2.0
Medium	0.25 to 0.50
Fine	0.125 to 0.25
Very Fine	0.0625 to 0.125

Porosity is the volume of void space in the sand. It is the volume of water needed to saturate a given weight of dry sand. Most sands have a porosity of about 30 to 40 % of the total volume. The finer a sand the greater its porosity. Permeability is the rate of flow or drainage of water through the sand. Fine sands have lower permeabilities due to their smaller pore sizes. Penetrability is related to particle size and porosity. It can be important to the macrofauna. All species must be able to burrow into the substratum. To determine the penetrability, the proportion of clay and silt and the water content are very important.

The two basic beach types are dissipative and reflective. Together with the intermediate types, there are six major microtidal beach types. Beach types can also be based on the degree of exposure. This ranges from very sheltered over sheltered and exposed to very exposed.

Ecological Features

Physical Environment

Sandy beaches, defined by their sand, wave and tidal regimes, range from narrow and steep (reflective) to wide and flat (dissipative), as sand becomes finer and waves and tides larger; most beaches are intermediate between these extremes. Reflective beaches are coarse-grained and have no surf zones, whereas dissipative beaches have finer sediments and extensive surf zones. Filtration volumes are higher on permeable reflective beaches, mainly driven by wave action, and lower on dissipative beaches, where tidal action drives most water throughput. Thus filtration is rapid on reflective beaches and slower on dissipative beaches, but the sand body of all open beaches is well flushed and oxygenated; only under conditions of very fine sand, such as on some dissipative and low energy beaches, can reducing conditions develop in the deeper sediment layers.

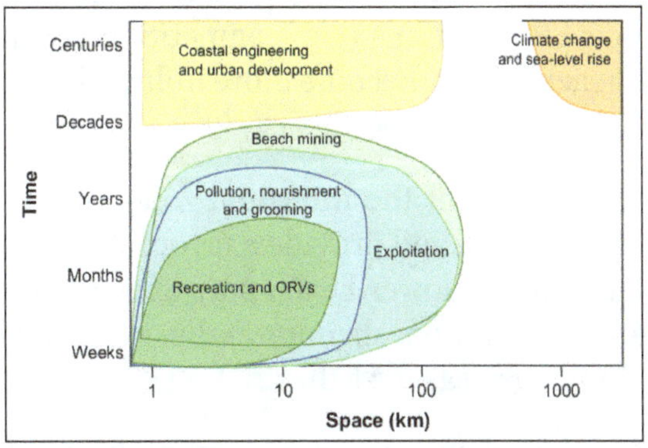

Conceptual model and schematic diagram showing the relative spatio-temporal scales in which different impacts reviewed here generally operate on sandy beach macrofaunal communities. Boxes/envelopes indicate the potential extent of individual impacts in space and time with the lower curve reflecting the lower limit of impacts in time and space, whereas the upper curve reflects the corresponding maximum. Shorter term impacts (i.e., weeks to months) tend to be pulse disturbances and effects are generally expected to last for shorter time periods, since sandy-beach species are adapted to severe physical disturbances (e.g., storms). However, the temporal extent of impacts from the anthropogenic pressures depicted here could be drastically altered if the intensity of the disturbance is increased and its timing is more protracted. Under such circumstances, sandy beach habitats could become unsuitable for supporting macro faunal communities and the ecosystem services they provide in the medium or long term.

Beaches are closely linked to surf zones and to coastal dunes through the storage, transport and exchange of sand; therefore impacts on beaches have consequences for these adjacent habitats. Sand transport, driven by waves on the west side and wind on the dry side, is highest in exposed surf zones, whereas sand storage is often greatest in well-developed dunes. Sand tends to move rapidly seawards across the beach and surf zone during storms and to return more slowly landwards during calms. In this way, storm wave energy is dissipated and the soft coast is protected from permanent erosion.

Faunal Components and Ecosystem Properties

The intertidal areas of beaches provide habitat for a diversity of fauna. The lacunar environment between the grains harbours interstitial organisms (bacteria, protozoans and meiofauna), forming a distinct food web. Larger macrobenthic invertebrates burrow actively and include representatives of many phyla, but crustaceans, molluscs and polychaete worms are usually dominant and encompass predators, scavengers, filter and deposit feeders. These macrobenthic invertebrates can reach high abundance (ca. 100,000 ind m^{-1}) and biomass (>1000 g m^{-1}), particularly in dissipative to intermediate beach types in temperate zones. Beaches that receive significant inputs of algae/seagrass wrack support a rich supralittoral fauna of crustaceans and insects. Most beach species are found in no other environment, their unique adaptations for life in these dynamic systems include: mobility, burrowing ability, protective exoskeletons, rhythmic behaviour, orientation mechanisms and behavioural plasticity.

The composition and abundance of invertebrate assemblages are controlled primarily by the physical environment, intertidal swash and sand conditions, being harshest on reflective beaches and more benign on dissipative beaches. Consequently, more species can colonise dissipative beaches, but fewer species, mainly robust crustaceans, can establish populations on reflective beaches. Whereas the effects of biological interactions (e.g. competition, predation) are overshadowed by physical factors on reflective beaches, they become more influential in structuring communities on dissipative beaches.

Faunal patterns change on either side of the intertidal zone. Supralittoral zones are important nesting areas for turtles and shorebirds, and provide a favourable habitat for invertebrates on stable reflective beaches. The fauna of the lower beach may extend their distribution seawards into the turbulent surf zone, where zooplankton, shrimps and prawns can be abundant; surf zones are also important nursery and foraging areas for fishes.

Food webs of sandy beaches are based on marine sources, such as phytoplankton, stranded algae, seagrasses and carrion. In dissipative systems, high productivity may be driven by surf phytoplankton and microorganisms, supporting benthic macrofauna and zooplankton as primary consumers and fishes and birds as top predators. While filtering water, the porous sand body and its biota mineralise organic matter and recycle nutrients, making beach ecosystems a crucial element in the nearshore processing of organic matter and nutrients.

Not only are beaches coupled to marine systems trophically, but they also interact physically and biologically with coastal dunes. Besides sediments, beaches and dunes exchange a variety of organic materials, and animals from both habitats move across the dune/beach interface to feed. Finally, beaches link terrestrial dune aquifers with coastal seas through the discharge of nutrient-rich groundwater.

Ecosystem Services and Sensitive Features

Sandy shores provide a wide range of ecosystem services, many of which are essential to support human uses of sandy coasts. The most important ecosystem services include: (1) sediment storage and transport; (2) wave dissipation and associated buffering against extreme events (storms, tsunamis); (3) dynamic response to sea-level rise (within limits); (4) breakdown of organic materials and pollutants; (5) water filtration and purification; (6) nutrient mineralisation and recycling; (7) water storage in dune aquifers and groundwater discharge through beaches; (8) maintenance of biodiversity and genetic resources; (9) nursery areas for juvenile fishes; (10) nesting sites for turtles and shorebirds, and rookeries for pinnipeds; (11) prey resources for birds and terrestrial wildlife; (12) provision of scenic vistas and recreational opportunities; (13) supply of bait and food organisms; and (14) functional links between terrestrial and marine environments in the coastal zone.

Certain features and processes on sandy shores are particularly vulnerable to impacts or change: (1) large volumes of sand are transported under high-energy conditions, so the consequences of disturbing sand budgets are most severe in such situations; (2) under low-energy conditions, water circulation is limited and therefore dispersal of materials, such as pollutants, is slow; (3) foredune vegetation is easily disturbed with resultant destabilisation of dunes; (4) because many beaches rely on marine organic subsidies, removal of such inputs (e.g. beach grooming) has serious consequences for beach food webs; (5) vertebrates that nest on the backshore are considered especially important in coastal conservation, and both turtles and shorebirds are easily disturbed in their nesting areas.

Functioning and Adaptations

The intertidal beach zone is covered part of the day by water and is part of the day exposed to air. Swash and tides supply nutrients and food. When the tide retreats, waste products, eggs and larvae are taken away. Organisms living on sandy shores have adapted to this dynamic environment.

Burrowing on high-energy sandy beaches must be rapid and effective. This is because the animals must not be swept away by uprushing and downrushing water. In contrast with rocky shores, desiccation is not an overriding concern, because the animals can retreat into the substratum or below the water table. Intertidal filter-feeders cannot feed while the tide has retreated. Many species of the meiofauna use vertical tidal migrations through the sand column. Other species move up and down the beach with the tides. There is a difference between directional stimuli (such as light, slope of the beach, water currents) and nondirectional stimuli (such as disturbance of the sand, changes in temperature, hydrostatic pressure). Directional stimuli act as orientational signs, while nondirectional stimuli act as releasing factors. The predominant feeding types are filter-feeding and scavenging. Animals on low-energy sandy beaches adapt their respiration differently from those on surf-swept beaches. Some adaptations are an increased ventilation rate, an increased ventilation efficiency, reduced metabolic rate or other ways of energy economy. Many sheltered-shore animals are facultative anaerobes as an adaptation to ebb tides. Other animals in oxygenated surf-swept beaches are essentially aerobic. The majority of the intertidal animals have a high tolerance to variability in their environment, even exceeding what is necessary for survival in their particular habitats. Some species bury themselves to escape high temperatures; others cool by evaporation, by entering the sea or by absorbing water from the substratum. Another problem for intertidal animals is the time of reproduction. There is variation in the number of eggs, the anatomy of the reproductive organs, the morphology of egg shells, times of breeding, mating behavior and developmental stages. Some species adapt by reproducing frequently (iteroparous) or by reproducing just once in a year (semelparous). There are also species that follow the lunar cycle to reproduce at the right time. To avoid predation, several behaviors have developed. The first one is deep burrowing. Another one is migration with the tide to escape predation. Crabs impress predators by holding their chelae open and aloft. According to circumstances, animals can modify their behavior. This is called phenotypic plasticity.

Several groups of vertebrates make use of sandy beaches for foraging, nesting and breeding. Turtles nest on the backshore of sandy beaches. Birds use the beach for foraging, nesting and roosting. Seals use several areas of the beach for nesting, molting, breeding and raising pups. Other terrestrial animals such as otters, baboons, raccoons, lions may descend onto the beach to forage.

Biota

The distribution and abundance of the sediment infauna is mostly controlled by complex interactions between the physicochemical and biological properties of the sediment.

The physicochemical properties are:

Grain size

- Water content
- Flushing rate of water through the sediment
- Oxidation-reduction state
- Dissolved oxygen
- Temperature
- Light
- Organic content

The biological properties are:

- Food availability and feeding activity
- Reproductive effects on dispersal and settlement
- Behavior that induces movement and aggregation
- Intraspecific competition
- Interspecific competition and competitive exclusion
- Predation effects

Most invertebrate phyla are represented on sandy beaches, either as interstitial forms or as members of the macrofauna. The macrofaunal forms are by far the better known. Some of them are typical of intertidal beaches and the surf zone, while others are more characteristic of sheltered sandbanks, sandy muds or estuaries and are less common on open beaches of pure sand.

Macrofauna

Macrofauna of sandy beaches are often abundant and, in some cases, attain exceptionally high densities. Their main feature is the high degree of mobility displayed by all species. These animals may vary from a few mm to 20 cm in length. The macrofauna community consists of organisms too large to move between the sand grains. The macrofauna of sandy beaches includes most major invertebrate taxa although it has been recognised that molluscs, crustaceans and polychaetes are the most important. There is a tendency for crustaceans to be more abundant on tropical sandy beaches or more exposed beaches and molluscs to be more abundant on less exposed and on temperate beaches although there are many exceptions of this and polychaetes are sometimes more abundant than either of these taxa. Generally crustaceans dominate the sands towards the upper tidal level and molluscs the lower down level . Physical factors, primary wave action and particle size of the sand largely determine distribution and diversity of the invertebrate macrofauna of sandy beaches. Food input and surf-zone productivity may determinate the population abundance.

Meiofauna

In contrast to the wave-swept beach surface inhabited by most of the macrofauna, the interstitial system is truly three-dimensional, often having great vertical extent in the sand. The porous system averages about 40% of the total sediment volume. Its inhabitants include small metazoans forming the meiofauna, protozoans, bacteria and diatoms. The meiofauna is defined as those metazoan animals passing undamaged though 0.5 to 1.0 mm sieves and trapped on 30 mm screens. On most beaches the interstitial fauna is rich and diverse, even exceeding in some cases the macrofauna in biomass . The dominant taxa of sandy beach meiofauna are nematodes and harpacticoid copepod with other important groups including turbellarians, oligochaetes, gastrotrichs, ostracods and tardigdades. See Latitudinal biodiversity patterns of meiofauna from sandy littoral beaches for a more detailed description of the latitudinal biodiversity patterns of meiofauna on sandy beaches.

Insects

Terrestrial insects and vertebrates are frequently ignored in accounts of sandy beaches. These animals are usually a conspicuous component of the ecosystems, often rivalling the aquatic macrofauna in terms of biomass and having a significant impact on the system with regard to predation and scavenging.

Threats to Sandy Beach Ecosystems

The accelerating destruction of natural habitats and consumption of natural resources by rapidly expanding human populations has caused huge impacts to ecosystems across the globe. Amplified by demographic population shifts towards the coast, many of these impacts are focussed at the world's coastlines, which are dominated by sandy shores. Intense coastal development, the inevitable consequence of economic progress, has resulted in widespread modification of sandy beach ecosystems. Human changes to sandy shores began at least two centuries ago, and are predicted to intensify exponentially over the next few decades. Global climate change, particularly sea-level rise, has added a new dimension to worldwide modifications of shorelines. It has therefore become increasingly critical to understand how sandy beach ecosystems and the goods and services they provide will respond to this unprecedented environmental change.

Exposed sandy beaches are physically dynamic habitats, inhabited by specialised biotic assemblages that are structured mainly by physical forces. Building on this perspective, documenting biotic responses to modifications of the physical environment is a critical step in predicting the consequences of global coastal change for beach ecosystems. Evidence for ecological change in beach ecosystems, which are exposed to human pressures at scales and intensities unmatched in history, is accumulating worldwide.

Here we provide a brief synopsis of the unique physical and ecological attributes of sandy beach ecosystems and review the main anthropogenic pressures acting on the world's single largest type of open shoreline. Threats to beaches arise from a range of stressors that span a spectrum of impact scales from localised effects (e.g. trampling) to a truly global reach (e.g. sea-level rise). These pressures act at multiple temporal and spatial scales, translating into ecological impacts that are manifested across several dimensions in time and space so that today almost every beach on every coastline is threatened by human activities.

Stressors

Recreation

Recreational seashore activities are overwhelmingly concentrated on sandy beaches. The effects of these pressures are particularly noticeable at scales ranging from weeks to months and from <1-100km Burgeoning coastal populations, coupled with more leisure time and improved mobility, have escalated the intensity and spatial ambit of beach recreation over recent decades. Being the prime sites for human recreation, beaches underpin many coastal economies. Beach management therefore customarily focuses on maximising the recreational experience for beach users, which often results in ecologically harmful human interventions such as nourishment, beach grooming, coastal armouring, and destruction of dunes to construct tourism infrastructure, and light and sound pollution.

Impacts caused directly by recreational activities are emerging as significant environmental issues. Dune vegetation is vulnerable to mechanical impacts caused by trampling, and modern beach management practices progressively seek to restrict human access to these sensitive areas. Nevertheless, camping and driving activities, which severely impact dune vegetation, both continue unabated in many parts of the world.

Evidence is sparse about how sensitive intertidal invertebrates might be to human trampling. Direct crushing of individual organisms on the unvegetated beach has been documented, and macrobenthic populations and communities respond negatively to increased human activity levels, but not in all cases. It can be difficult to separate the effect of human trampling from habitat modifications (e.g. seawalls replacing foredunes) because these often coincide in high-use areas.

Impacts of trampling on supralittoral fauna are usually seasonal at medium-high latitudes, enabling some recovery during periods of low use. However, press disturbances are expected on low-latitude beaches where the intensity of the disturbance is much higher than in temperate areas and its timing is more protracted. Under such circumstances, beach habitats could become unsuitable for sandy beach macrofauna in the medium or long term. It is therefore necessary to establish the ecological carrying capacity of beaches in terms of direct use, and to develop mechanisms for controlling access, tasks that many authorities neglect.

Beaches and diunes are critical habitats for shorebirds and turtles and both taxa are sensitive to disturbances. For example, human activities disturb shorebirds, modifying key behavioural traits that are crucial to their survival and reproduction, including: (1) changes to foraging behaviour resulting in less feeding time, temporal shifts in feeding times and decreased food intake; (2) decreased parental care when disturbed birds spend less time attending the nest, thus increasing exposure and vulnerability of eggs and chicks to predators; and (3) decreased nesting densities in disturbed areas and population shifts to less impacted sites.

Off-road vehicles (ORVs) are commonly used on beaches and dunes worldwide and cause damage that includes: (1) disturbing the physical attributes and stability of dunes and beaches by deeply rutting the sand surface and destroying embryonic foredunes in the tyre tracks; (2) destroying dune vegetation, leading to lower diversity and less floral ground cover; and (3) disturbing, injuring or killing beach fauna, including endangered vertebrates such as turtles and shorebirds. Whether direct mortality of beach invertebrates caused by ORVs propagates to higher-order ecological effects, such as disruptions of food web linkages to fishes and raptors, is currently unresolved. Nevertheless, positive population responses by shorebirds following exclusion of ORVs suggest that ORV-impacts can extend beyond the individual organism level of ecological organisation.

Cleaning

Cleaning or grooming, a common practice on beaches heavily used for tourism, clears beaches of macrophyte wrack, litter and other debris by raking and sieving the sand, often with heavy equipment. Grooming removes not only unwanted material, but also propagules of dune plants and other species, and it perturbs resident organisms and roughens the sand, thereby exposing a greater surface area to the erosive effects of wind.

Wrack removal has significant ecological consequences, especially in regions with high levels of marine macrophyte production. The community structure of sandy beach macroinvertebrates can be closely linked to wrack deposits, which provide both a food source and a microhabitat refuge against desiccation. Wrack-associated fauna, such as talitrid amphipods, oniscoid isopods and insects, which can comprise up to 40% of the species and represent important prey for higher trophic levels, are significantly reduced in species richness, abundance, and biomass by grooming. By contrast, taxa with well-developed dispersal abilities, such as flies, can be more prevalent on groomed beaches.

Upper intertidal meiofauna may also be impacted by grooming, although these effects can be confounded with those from trampling. Meiofauna communities may recover relatively quickly (24 h) from a single, short-term grooming event, but it remains unknown what the consequences are of repeated, regular beach cleaning activities. Effects of this stressor could be noticeable at scales ranging from weeks to years and from < 1 to 100 km.

Because shorebird numbers are positively correlated with wrack cover and the biomass of their invertebrate prey that feed on wrack, grooming will lower bird numbers. The heavy equipment used in beach grooming can also cause direct mortality of the eggs and young of beach-nesting shorebirds, turtles and fish; many groomed beaches therefore no longer support breeding populations of these vertebrates. Furthermore, tracks created by grooming can disorientate turtle hatchlings trying to reach the sea. Finally, grooming can result in abnormally broad unvegetated zones that are inhospitable to dune formation or plant colonisation, thereby enhancing the likelihood of erosion.

Invertebrates may recover slowly from the effects of grooming, especially if it is conducted on a daily or weekly basis throughout the year, as is the case along many developed coastlines. For example, nearly half (>160 km) of the sandy beaches in southern California are regularly groomed, and some beaches in Los Angeles are groomed twice a day. Although beach cleaning is generally undesirable from an environmental conservation perspective, simple mitigation strategies could entail alternating areas of natural (uncleaned) beach with groomed sections, if grooming is deemed essential to provide clean beaches for recreation.

Nourishment

More than 70% of the world's beaches are experiencing erosion. Because engineering solutions, such as seawalls, breakwaters and groynes, are often ineffective to the point of causing the loss of the intertidal beach, beach nourishment (also called beach replenishment, restoration or renourishment), has increasingly been used to combat shoreline erosion. Nourishment is preferred on both economic and conservation grounds, but it can cause ecological damage to sandy beach habitats and biota. The typical scale of nourishment is at 1 –10 km and at temporal scales ranging from weeks to years. Impacts occur both at sites from which sediment is extracted and at the receiving environment. These impacts manifest at the population (demography and dynamics), community (species richness) and ecosystem (functional processes, nutrient flux, trophic dynamics) levels. Affected biota includes benthic micro-algae, vascular plants, terrestrial arthropods, intertidal and subtidal invertebrates, other marine zoobenthos and avifauna.

Factors influencing the nature and extent of ecological impacts of nourishment include the mechanical process itself, its timing, and the quality and quantity of new sediment placed. Effects may be direct, such as mortality of organisms when buried, or indirect, such as reduced prey availability for shorebirds. Most research has targeted the effects of altered sediment quantity and quality on macrofauna. The immediate impacts are usually large and may be caused either by burial or by emigration. These effects may be compounded by changes in beach morphology, particularly when nourishment creates a steeper beach and reduces the habitat area for some species. Nourishment can also disturb the nesting and foraging of birds, destroy dune vegetation and compact the sand. Compaction affects the interstitial spaces, capillarity, water retention, permeability and the exchange of gases and nutrients.

Nourishment of sandy beaches usually acts as a short-term, pulse disturbance that elicits a pulse ecological response (i.e. recovery occurs). This is expected since sandy-beach species are adapted to severe physical disturbances, storm events having been a frequent feature of their evolutionary history (Hall, 1994). Recovery probably occurs in months rather than years, but the trajectory depends on sediment quality. However, if the profile of the nourished beach and the imported sediments do not match the original conditions, recovery of the benthos is unlikely. Thus, where unnaturally coarse or fine sediments are used severe ecological impacts may occur and recovery is protracted. In addition, artificially flattened and extended sand bodies can be colonised by rapidly moving opportunistic macrofauna; under these conditions, few species dominate and biodiversity is reduced.

Mitigation of ecological impacts of nourishment is often impeded by limited data about the life history of the affected species, recovery rates and the cumulative effects of repeated nourishment events. Nevertheless, basic management recommendations include: (1) the avoidance of sediment compaction; (2) careful timing of operations to minimise biotic impacts and enhance recovery; (3) the selection of locally appropriate techniques; (4) the implementation of several small projects rather than a single large project, including repeated application of sediment in shallow layers (<30 cm) rather than single pulses that kill the fauna by deep burial; (5) interspersion of nourished beach sections with unaffected areas; and (6) importing sediments and creating beach profiles that match the original beach conditions as closely as possible.

Pollution

Pollution is a sensitive issue on beaches given their immense value for recreation and tourism. Pollutants act at a variety of spatial and temporal scales and include a variety of anthropogenic materials, ranging in size from molecules to large debris, and can impair the physiology, survival, reproduction and behaviour of species in all habitats of the beach from interstitial environments to the surf zone. Pollution can also cause aesthetic disturbances, thereby impacting tourism industries that depend on public perceptions of clean beaches.

Most solid waste stranded on the shore is allochthonous, brought ashore by waves and currents. Plastic is extremely persistent and it dominates the visible litter on sandy beaches worldwide. The major dangers associated with plastic include ingestion by and entanglement of vertebrates, such as seals, seabirds and turtles. In terms of human health, risk increases when medical plastic wastes reach the shore from coastal dumping sites. Economic losses can also arise when tourist beaches are persistently contaminated by litter from land or ocean sources.

Wastewater and sewage are significant sources of beach pollution. Intertidal sands and surf-zone waters can be contaminated by pathogens, including bacteria and filamentous fungi, which are delivered into the system by sewage systems discharging directly into coastal waters or estuaries near beaches. Bacterial levels in surf zone waters that exceed human health standards are a frequent cause of beach closures and public warnings on many developed coasts. Contamination at any given beach is dependent on: (a) the appropriateness of waste-water management practices; (b) the timing and intensity of local rainfall events and subsequent runoff; and (c) the strength of mixing and dispersion in the surf zone. Metal pollution from wastewater and industry accumulates preferentially on fine-grained beaches, which usually supports relatively high diversity and biomass, thus causing significant reductions both in biodiversity and in the population density of economically valuable species.

Freshwater effluents arising from human activities can also cause a broader deterioration in the quality of the surrounding habitat, with impacts at various levels of ecological organisation, ranging from individuals and populations, which might experience reduced survival, growth and fecundity rates, through to the community, which can exhibit changes in diversity and structure.

Oil spills are potentially the most destructive pollution source impacting sandy beaches, affecting all trophic levels. Impacts can be acute and temporary, but they can also be more chronic, lasting for many months or even years. Beach morphodynamics and exposure strongly influence the duration of contamination: the coarser the sediment (i.e., reflective beaches), the more rapidly and deeply oil penetrates, sometimes even reaching below the groundwater table. Sheltered beaches are generally more sensitive to pollution than exposed beaches, even when sediments are fine and oil does not penetrate deeply, because they are less well flushed by wave action and subsurface oil persists much longer than surface oil. Persistence and breakdown of stranded oil depends on particle size, wave energy, temperature and other factors, including fungal degradation Q5. Methods used for oil spill cleanup on beaches may also cause ecological impacts. Public concern about oil spills has focused on catastrophes like the IXTOC 1 in Mexico, the Exxon Valdez in Alaska and the Prestige in Spain and many smaller events tend to go unreported, but introduce considerable quantities of oil into the beach environment. Tar balls that originate from the (often illegal) flushing of ships' bilges at sea are present on most beaches, especially near oil shipping routes.

In most cases, management measures against pollution have to be taken without a preceding environmental impact assessment, as has occurred been the case of oil spills affecting beaches all over the world. Oil pollution impacts differ between chronic cases and episodic events. The former seldom require direct action, whereas the latter may need a major cleanup, which in turn causes impacts. Management requires monitoring and control of water and sediment quality, particularly in urban areas.

Exploitation

Artisanal invertebrate fisheries are the commonest form of exploitation on sandy beaches. Although these are generally of reasonably small scale, for various reasons impacts can be significant. First, target species tend to occur in patches, and fishers can therefore easily target and serially deplete dense aggregations. Most exploited sandy beach stocks are also short lived and susceptible to recruitment overfishing in the short term. Moreover, harvesting activities cause incidental mortalities, both directly through physical damage of organisms and indirectly when sediment disturbance lowers habitat quality and suitability. Such incidental mortality is not limited to the exploited fraction of the target populations, but also affects their unexploited fractions and non-target species on fished beaches.

Long-term, large-scale field experiments have demonstrated a predictable pattern of responses to fishing: elevated harvesting rates significantly reduce recruitment to the exploitable stock and to the population as a whole, resulting in a declining spawning stock, a decreasing age/size at maturity and a decreasing proportion of older individuals in the catch. Several fishery-related density-dependent processes have also been identified. For example, fishery closures can result in elevated adult densities, which cause density-dependent increases in natural mortality rates of older clams, as well as decreases in growth and survival rates of youngof the-year, the inhibition of recruitment of both target and sympatric species, and the reduction of age-specific fertility.

Where fishing-related disturbance is a pulse of limited spatial extent, its impacts can be ameliorated relatively quickly. However, chronic disturbance can cause large-scale and persistent impacts for populations and communities, as documented for Argentinean beaches where fishing involving the use of tractors caused lasting damage to populations of yellow clam Mesodesma mactroides for decades. Detrimental effects of fishing can also be exacerbated when they are superimposed on environmental change, irrespective of whether such change is of anthropogenic origin, or not. On a New Zealand beach, clam populations depleted by artisanal fisheries between the mid-1960s and 1990 failed to recover following the closure of the fishery, apparently because continuous erosion of beaches has dramatically reduced the clam habitat. Similarly, the effects of fishing can be amplified or reinforced by mass mortalities associated with blooms of toxic algae, by parasitism and by temperature anomalies.

Technological interdependencies among user groups and ecological interdependencies between stocks are common in sandy beach fisheries. Even fisheries in which juveniles and adults are spatially segregated are not immune to this phenomenon, with separate user groups often targeting intertidal and subtidal stocks using different gear types. Moreover, commercial fishers and recreational anglers, who collect large specimens for bait, tend to be more selective than recreational food collectors. In this way, different user groups, which are often difficult to distinguish, can inflict different patterns of fishing mortality on the resource. Thus, beach fisheries tend to involve a complex blend of users, requiring the explicit incorporation of cultural, social and political dimensions in resource management. As harvesting not only affects the targeted species directly through fishing mortality, but it also has collateral impacts that can alter ecosystem structure and functioning, sandy beach fisheries can have impacts disproportionate to their economic value. Being vulnerable to fishing, few beach fisheries have scope for growth. Because it is difficult to restrict participation in these accessible fisheries, management failure and stock depletion seem to be the norm rather than the exception.

Biological Invasions

Human activities that are vectors for species introductions to beaches are not a recent phenomenon. Among the oldest-known human agents of species introduction to beaches is the practice of exchanging large quantities of dry ballast sand and gravel on shore during ship maintenance, which dates back to the early days of ocean-going sailing vessels.

Nearshore and intertidal benthic habitats in many regions have been invaded by alien macrophytes that displace native algae and sea grasses. Although not invaders of beaches in the sense of establishing there, they can form large wrack deposits on sandy beaches when detached from their benthic habitats. In Hawaii, accumulations of the invasive Hypnea musciformis of >9000 kg week 1 have been reported for pocket beaches. In Argentine Patagonia, the biomass and composition of macrophyte wrack on beaches have been changed by the invasive kelp Undaria pinnatifida. The kelp Sargassum mutica is accumulating in large quantities on beaches of northwestern Spain (Lastra, personal observation), and the green algae Caulerpa taxifolia can replace the native seagrass Posidonia oceanica as the main component of wrack on Mediterranean beaches. The preference of wrack consumers (e.g., talitrid amphipods) for macroalgae suggests the potential for impacts of invasive algae on food webs and nutrient cycling on beaches.

Few invasive invertebrates have been reported from sandy beaches to date, but this may reflect poor sampling coverage rather than a lack of biological invasions. The semi-terrestrial talitrid amphipod Orchestia cavimana, originally from freshwater habitats in the Black Sea and Mediterranean region, has been reported from fresh to brackish habitats in the British Isles, and more recently has been spreading rapidly on open exposed shores with low salinities of the NE Baltic Sea. The talitrid Platorchestia platensis, which presently inhabits Europe, the Canaries, Japan, India, Hawaii and other Pacific islands, and which was recently reported in the Baltic, may be an exotic species. A few widespread polychaetes (e.g., glycerids and spionids) reported from exposed beaches, are classified as cryptogenic, implying that their origin is unknown and that they are neither demonstrably native nor introduced. The best known of these are Glycera americana and Scololepis squamata, both of which are reported from beaches and estuaries of several continents. The possibility that some cryptogenic forms could represent cryptic species rather than a single widespread invasive species needs to be considered.

Invasive terrestrial insects can also impact beach ecosystems. Red fire ants (Solenopsis invicta) from Brazil, are now found in the US A, Australia, Taiwan and China; they can prey on the eggs and hatchlings of loggerhead and green sea turtles. A congener, the black fire ant (Solenopsis richteri) preys intensely on intertidal polychaetes in Argentinean lagoons, which suggests that it could also consume intertidal invertebrates on exposed beaches. The Argentine ant (Linepithema humile , formerly Iridomyrmex humilis), a small, dark ant native to northern Argentina, Uruguay, Paraguay, and southern Brazil, has invaded a variety of habitats, including beaches, in South Africa, New Zealand, Japan, Australia, Europe and the U. S.

Coastal Development and Engineering

The expanding use of coasts by human populations characteristically involves a gradual intensification of urban development in the littoral-active zone, with the ultimate consequence that coupled surf–beach –dune systems must be managed. This management most commonly focuses on the sediment budget. Modern coastlines are increasingly starved of sand as dams trap sediments that would otherwise feed beaches; the sediment budget is further disrupted by activities such as quarrying, land reclamation, urbanisation, afforestation and agricultural use. As a result, most modern coastlines are experiencing accelerating rates of erosion.

Society's response to beach erosion and shoreline retreat relies heavily on engineering interventions that place armouring structures on beaches. Hard structures, such as walls constructed of stone, concrete, wood, steel or geotextiles have been used for centuries as a coastal defence strategy, but this protection is not achieved without ecological costs. Armouring structures alter the natural hydrodynamic system of waves and currents, thereby affecting sand transport rates, which in turn control the erosionaccretion dynamics of beaches. Intertidal seawalls and other structures that reflect wave energy and constrain the natural landward migration of the shoreline have unplanned environmental impacts, such as flanking erosion of shorelines adjacent to those protected by engineering structures. Passive erosion occurs on the armoured beaches themselves; and because seawalls arrest the landward retreat of the shoreline in the face of erosion, the beach seawards of the structure is frequently drowned. Armouring might also enhance beach erosion on protected coasts, although this remains controversial.

The impacts of seawalls and other coastal armouring structures may cause significant habitat changes, with attendant ecological impacts that can be difficult to detect in the short term. As eroding beaches become narrower after armouring, the reduced habitat can directly lower the diversity and abundance of biota, especially in the upper intertidal zone. This, in turn, can also be detrimental to higher trophic levels, e.g., coastal avifauna may be impacted both by reduced habitat area and by declining intertidal prey resources. This phenomenon is reflected by observations of significantly lower numbers and fewer species of birds on armoured compared with unarmoured segments of Californian beaches. The loss of dry-sand following armouring also eliminates nesting habitats for turtles and some specialised species of fishes (e.g. California grunion, surf smelt in the USA).

Armouring may also affect the quality of beach habitats. For example, the rates of deposition and retention of macrophyte wrack, driftwood and other natural debris that can be important to beach biota as food and habitat, are lower on armoured than unarmoured beaches. By contrast, increased deposition of wrack can occur on some beaches where offshore structures have been placed. These offshore defences also alter physical characteristics and benthic communities of beaches and surf zones.

Because shore-protection structures often present novel habitats on beaches, they could promote the introduction of exotic species. Together, engineering structures and introduced species could severely impact native assemblages, not only by modifying selection pressures as a result of habitat transformation, but also by introducing competition. However, this issue remains speculative and more work is needed to assess its importance for sandy beach ecosystems.

Sea-level rise and other large-scale and long-lasting effects of global warming are expected to intensify beach erosion. To protect human assets on soft-sediment coastlines globally in the face of escalating erosion, the use of coastal armouring will therefore expand substantially. For the extensive armoured coastlines in many countries, changes in sea level will shift the locations of existing armouring structures to lower positions on the shore, amplifying interactions with waves and tides and further accelerating beach erosion. Thus, the combined effects of rising sea levels and coastal armouring are predicted to cause ecological impacts to beaches on unprecedented scales.

Mining

Beach sands have been mined in many places, including Sri Lanka for building lime from coral sand, Tanzania, Korea, Argentina, Italy and many other countries for building sand, the Namibian coast for diamonds, and several areas for heavy minerals such as titanium and zirconium. Further, mine tailings are discharged onto beaches in many places. Diamond mining activity in Namibia has been shown to negatively impact shorebirds, and tailings from copper mines cause lower meiofauna density and diversity on Chilean beaches. Removal of sand disturbs sediment budgets, possibly contributing to enhanced erosion; it may also alter particle size, changing the morph dynamic state of beaches and intertidal benthic communities. Thus, the general effects of any form of mining on beaches are to damage the beach and dune habitat, to alter the sediment budget and hasten erosion. It is now illegal in many countries to remove sand from the littoral active zone and the practice should in general be prohibited. In many places sand is mined offshore, often on the outer shelf; the ecological impacts of this activity are probably slight for beaches, but not well understood.

Climate Change

Although the exact magnitude of physical changes resulting from global climate change is still uncertain, ecological responses (e.g., changes in phenology, physiology, range and distribution, assemblage composition, species interactions) are increasingly apparent in sandy beaches. However, there are no direct studies of effects of this long-lasting and large-scale stressor on beach ecosystems; consequently, many of our predictions of the likely impacts on beach ecosystems are derived from other systems.

Rising temperatures may have different implications at different latitudes and for taxa with different dispersive abilities and ranges. Because many sandy beach species (e.g. peracarid crustaceans) lack dispersive larval stages, their rates of range extension might be outpaced by changes in temperature, making these taxa particularly vulnerable to climate-change effects. Narrow-range endemic species would be at greatest risk and some species could eventually be replaced by species from lower latitudes. But even migratory species, as well as those that have pelagic larval stages, may be impacted directly by large-scale, climate-driven changes to prevailing oceanographic systems.

Beach biota may also respond to indirect effects of temperature change on beaches. For example, comparatively small increases (wo.6 C) in temperature have been associated with major changes in planktonic ecosystems in the North Atlantic over the past five decades. Given that plankton is a key food source for suspension-feeding beach species, changes in plankton communities will have unpredictable impacts on sandy beach macrofauna. Changes may be significant for semi-terrestrial species (peracarids and insects), which will be affected both directly and indirectly by changes in water and air temperatures.

A direct consequence of warming seas is sea-level rise. Current estimates suggest that sea level is rising by an average of 1.7 ± 0.5 mm yr^{-1}, but there is potential for far greater rates of change. Irrespective, rising sea level pushes the high-water mark landward, causing beaches to migrate slowly inland. Low-gradient dissipative shores, which house the greatest biodiversity, are at most risk due to their erosive nature and the much greater run-up of swashes on gentle gradients. Moreover, warmer air and sea temperatures would translate into more frequent and more severe storms, thus escalating beach erosion to the point where entire beaches could disappear, removing habitat for the biota. The institution of setbacks is currently international best practice. This involves defining a shore normal zone within which new development is prohibited and space is thereby provided for natural coast retreat. Complementary to this is managed coastal retreat, which involves preparing to abandon existing infrastructure within the setback zone and/or developing plans to remove these existing structures that are too close to the shore. Soft solutions, such as nourishment, should be sought first, with hard engineering the very last resort.

Predicted changes in ocean acidification could further impact sandy beaches. The continued reduction of pH in surface waters (0.1 unit less than pre-industrial levels and predicted to decline by a further 0.3 –0.4 units by the end of the century) will reduce calcification rates and calcium metabolism in marine organisms, including several sandy beach molluscs and crustaceans.

Management to maintain beach habitat requires long-term mitigation and adaptation strategies. Mitigation would arrest climate change and its consequences via large reductions in greenhouse emissions. Adaptation would provide setback zones, allowing the beach to migrate inland as the sea rises. The latter would have minimal ecological consequences for beaches, but would be very expensive in urban areas. Alternatively, if engineering solutions (e.g. seawalls) are used to defend societal assets, intertidal sand habitat will be lost. Adaptive measures accept the reality of sea-level rise and coastline retreat and seek to increase coastal resilience, a concept with ecological, morphological and socio-economic components. Unfortunately, coastal resilience has been impaired by anthropogenic effects on the sediment budgets of beach-dune systems. Measures to promote resilience include the protection, vegetation and stabilisation of dunes, the maintenance of sediment supply and the provision of buffer zones, rolling easements or setbacks that allow the landward migration of the coastline.

Salt Marsh Ecosystem

Salt marshes are coastal wetlands that are flooded and drained by salt water brought in by the tides. They are marshy because the soil may be composed of deep mud and peat. Peat is made of decomposing plant matter that is often several feet thick. Peat is waterlogged, root-filled, and very spongy. Because salt marshes are frequently submerged by the tides and contain a lot of decomposing plant material, oxygen levels in the peat can be extremely low—a condition called hypoxia. Hypoxia is caused by the growth of bacteria which produce the sulfurous rotten-egg smell that is often associated with marshes and mud flats.

Characteristics

Salt marshes typically develop on depositional coasts, bays, and estuaries where tidal action is relatively gentle and erosion no more than intermittent and light enough to allow vegetation to take hold. They are common on low-energy costs such as estuaries, enclosed bays, and the land sides of barrier islands and strips.

A grass salt marsh in the Marine Park Nature Center in Brooklyn

Plant species diversity is relatively low, since the flora must be tolerant of salt, complete or partial submersion, and anoxic mud substrate. The most common salt marsh plants are glassworts (Salicornia spp.) and the cord grasses (Spartina spp.), which have worldwide distribution. They are often the first plants to take hold in a mudflat and begin its ecological succession into a salt marsh. Their shoots lift the main flow of the tide above the mud surface, while their roots spread into the substrate and stabilize the sticky mud and carry oxygen into it so that other plants can establish themselves as well. Plants such as sea lavenders (Limonium spp.), plantains (Plantago spp.), and varied sedges and rushes grow once the mud has been vegetated by the pioneer species.

Saline tidal marshes that are covered and exposed daily by the tide may be covered by the tall form of smooth cordgrass (Spartina alterniflora), while areas covered by water only sporadically may be populated by short smooth cordgrass, spike grass, and black grass (Juncus gerardii).

The flora of a salt marsh is differentiated into levels according to the plants' individual tolerance of salinity and water table levels. Vegetation found in the water must be able to survive high salt concentrations, periodical submersion, and a certain amount of water movement, while plants further inland in the marsh can sometimes experience dry, low-nutrient conditions.

Salt marshes are quite photo synthetically active and are extremely productive habitats. They serve as depositories for a large amount of organic matter, and are full of decomposition, which feeds a broad food chain of organisms, from bacteria to mammals. Interestingly (in such a biologically productive biome), many of the halophytic plants such as cordgrass are not grazed at all by higher animals, but die off and decompose, to become food for microorganisms, which in turn become food for fish and, in turn, birds.

A salt marsh in the Marine Park Nature Center in Brooklyn

In wintertime in colder climates, the saltmarsh looks more open than in summer. More space is seen between reeds as snow falls between them. The water partly freezes, which makes it look rather spectacular. In warmer climates, salt can be produced by solar energy so that the French equivalent of the salt marsh, the marais salant has come to be envisaged primarily as an industrial plant, though classified as a form of agriculture, known in French as saliculture.

Features of Trends in Extent

Salt marshes are among the most productive temperate ecosystems in the world. Contemporary salt marshes developed within the last 8,000 years in low-energy, coastal locations in response to rising sea levels. Their ecology and global importance has been described in classic literature such as Chapman, Ranwell, Doody and Adam. The physical stresses of salinity and flooding generate zones of salt-tolerant emergent vegetation, including such genera as Carex, Spartina, Juncus, Salicornia, Halimone, Puccinellia, and Phragmites. Marsh grasses contribute to the accumulation of organic matter and trapping of inorganic sediment. Salt marsh sustainability is mainly controlled by the relationship between marsh vertical accretion (due to sediment accretion, peat accumulation, belowground decomposition, subsidence) and sea level rise. Other facts that impact their development and structure are tidal, wave and current action, erosion, freshwater influx, nutrient supply, and topography. Less than 50 percent of the world's original wetlands remain and current loss is estimated at 1-2 percent per year making wetlands one of the fastest disappearing ecosystems worldwide. Salt marsh loss coincides with a general historical degradation of estuarine ecosystems. Just upstream from salt marshes are brackish marshes with lower salinity regimes, which are also highly productive, subject to the same stresses, and of equally great conservation concern.

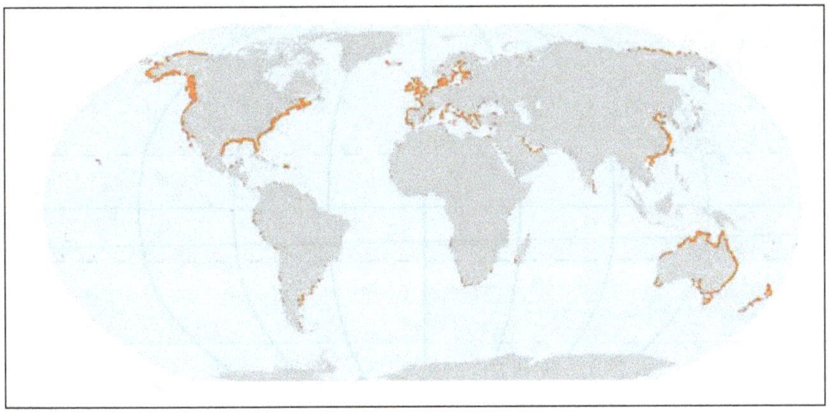

Salt Marshes (in orange)

Major Pressures Linked to the Trends

Over 60 percent of the globe's population lives on or near the coast, and coastal populations are increasing at twice the average rate, making coastlines highly vulnerable to human activities. Salt and brackish marshes, formerly viewed as useless wastelands, were filled in for urban or agricultural development. Reclamation of land for agriculture by converting marshland to upland was historically a common practice. Coastal cities worldwide have expanded onto former salt marshes and used marshes for waste disposal sites. Airoldi and Beck estimate that countries in Europe have lost over 50 percent of their salt marsh and seagrass areas to coastal development. Estuarine pollution from organic, inorganic, and toxic substances is a worldwide problem. Marshes have been drained, diked, ditched, grazed and harvested. They have been sprayed for mosquito control, and have been invaded by a range of non-native species that have altered their ecology. As one example, Massachusetts, United States of America, has lost 41 percent of its salt marshes since the 1770s, with a loss of 81 percent in Boston.

Key threats to salt marshes are land reclamation, coastal development, dredging, sealevel rise (SLR), hydromodification, alteration of processes (e.g. sediment delivery, freshwater input) and eutrophication. Accelerated SLR is the largest climate-related threat to salt marshes. The Intergovernmental Panel on Climate Change predicts with medium confidence a SLR of 0.26-0.98 m by 2100. Nicholls et al. predict that 1 m SLR will eliminate 46 percent of the world's coastal wetlands. Some salt marshes can keep pace with SLR, but others, especially those cut off from their sediment delivery via levees and seawalls cannot. Nutrient pollution, which destabilizes below-ground biomass and increase decomposition, is likely to exacerbate salt marsh loss due to SLR. Subsidence, which contributes to relative SLR in some regions, is an additional stressor. The impact of SLR will depend upon accretion and subsidence rates and other processes that influence the marshes ability to grow vertically and/or to migrate inland. "Coastal squeeze" describes the limitation of marshes to extend landward due to boundaries such as paved areas, seawalls, and bulkheads from coastal development.

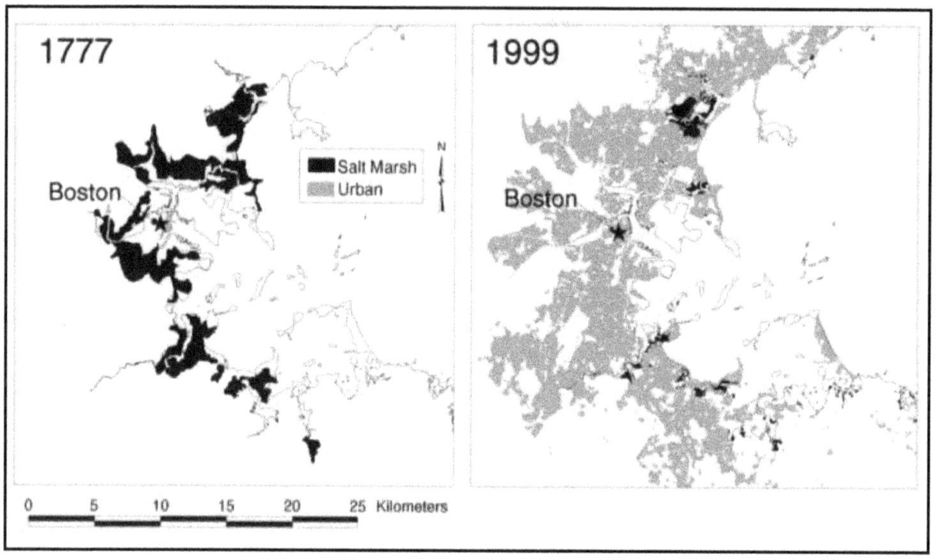

Salt marsh loss in Boston, Massachusetts

Implications for Services to Ecosystems and Humanity

Salt marshes play a large role in the aquatic food web and the cycling of nutrients in coastal waters. They serve as critical habitat for various life stages of coastal fisheries that account for a large percentage of the world's fish catch. Over half of the commercial fish species of the East coast of the United States utilize salt marshes at some time of their lives. In addition to providing habitat for juvenile fishes, crabs, and shrimps, marshes support populations of some small forage fishes, which come up on the marsh surface at high tide to feed on invertebrates. Many migratory shore birds and ducks use salt marshes as stopovers during migrations and some birds winter in the marsh. Wading birds, such as egrets and herons, feed in salt marshes during the summer. Continued marsh loss could therefore dramatically alter estuarine food webs.

SLR is increasing the vulnerability of coastal populations to coastal erosion, flooding, and storms. Salt marshes serve as natural barriers to these coastal hazards. They serve as shoreline stabilizers because they attenuate wave energy and help prevent erosion. They also slow and store floodwaters, reducing storm impacts on coastal communities. While wetlands do not provide complete protection against coastal hazards, even small salt marshes can provide significant shoreline protection. Their preservation and restoration may significantly decrease the economic impact and human losses of extreme events such as hurricanes and tsunamis.

Salt marshes remove sediment, nutrients, microbes, and contaminants from runoff and riverine discharge, acting as sponges absorbing much of the runoff after major storms and reducing flooding. They sequester pollutants from the water that drains down from the land, protecting nearby estuarine areas and coastal waters from harmful effects. They play a major role in the global carbon cycle and represent a major portion of the terrestrial biological carbon pool. They store excess carbon in their sediments, preventing it from re-entering the Earth's atmosphere and contributing to global warming. Salt marshes are thus an important component of the world's "blue carbon" and currently are being incorporated into global carbon markets. Chmura et al. estimated that tidal wetlands sequester 10 times the amount of carbon sequestered by peatlands. Salt marshes also provide excellent tourism, education, and recreation services, as well as research opportunities.

It is clear that salt marshes provide enormous benefits to society in the form of "ecosystem services". In this regard, coastal wetlands (which include salt marshes) are among the highest valued coastal ecosystems. The serious reduction in salt marsh area reduces their capacity to provide these critical ecosystem services.

Importance

Historically, salt marshes have sometimes been treated as "wastelands," along with other wetlands. We now know that salt marshes are one of the most biologically productive habitats on the planet, rivaling tropical rainforests. This is partly due to the daily tidal surges that bring in nutrients, the natural chemical activity of salty (or brackish) water, the tendency of nutrients to settle in roots of the plants there, and the tendency of algae to bloom in the shallow unshaded water.

A walkway for people over a salt marsh

Salt marshes also provide a benefit by protecting against severe weather, such as in the North American Gulf Coast. Tidal marshes slow erosion along the shoreline and buffer stormy seas. They also help to filter pollutants before they enter oceans and estuaries, either by settling of sediments or microbial and plant removal of nutrients and other substances.

In the past, substantial areas of saltmarsh have been reclaimed as agricultural land and for urban development, but in the United States and Europe they are now accorded a high level of protection by the Clean Water Act and the Habitats Directive respectively. There is growing interest in restoring salt marshes, through a process of managed retreat.

Deep Sea Ecosystem

The deep sea floor represents the largest habitat on earth. It ranges from the edge of the continental shelf at 200m to the bottom of the ocean. At the edge of the continental shelf is the shelf break, where the gradient of the floor increases down the continental slope. Below the continental slope lies the continental rise, which has a more gently slope. At around 4,000 meters depth, the ocean floor is reached and extends over the ocean basins at depths of 5,000 metres on average. This is called the abyssal plain. The zone between the continental shelf and the abyssal plain is the bathyal zone. In some places, the sea floor drops again into elongated trenches with depths of 10 to 11 kilometres. This region is the hadal region. The ocean floor is interrupted by mountain chain known as the mid-oceanic ridge system. Other features on the ocean floor are seamounts and hydrothermal vents.

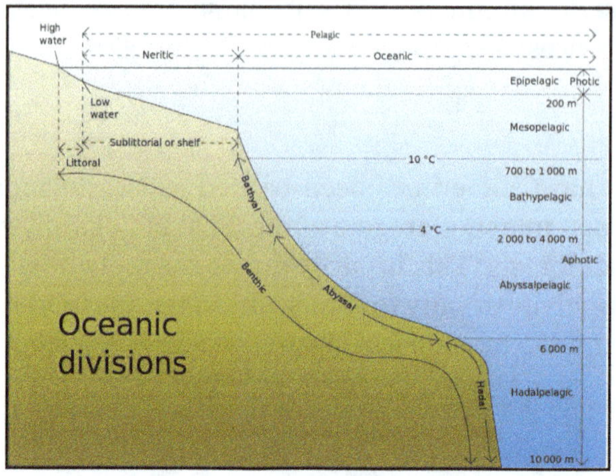

Ocean zones

Characteristics

The temperature of the waters of the deep sea varies from 4 °C to -1 °C. Exceptions are the Mediterranean Sea with an average temperature of 13 °C in the upper 4,000 metres and the Red Sea where the bottom water can reach temperatures up to 21.5 °C at depths of 2,000 metres. The lowest temperatures are found in the Antarctic Ocean and are about -1.9 °C. The salinity of the seawater is relatively constant. Below 2,000 metres, the salinity is close to 35 ‰ (34.8‰ ± 0.3 ‰) and at the very deepest layers it is 34.65‰. The oxygen concentrations are near saturation except in the oxygen minimum layer between 0.5 and 0.6 kilometres depth. In enclosed basins such as the Black Sea, the water is anoxic below 250 metres. When the deep water is further from its origin the oxygen concentration declines by metabolic processes. This is e.g. the case in the deep North Pacific. The most predictable physical variable is hydrostatic pressure. Per 10 metre the depth increases, the pressure increases by 1 atmosphere (10 Pa). In the deep ocean, light is almost absent. It is a dark environment and the organisms are adapted to it. The circulation in the deep sea is explained in the article ocean circulation. The deep sea bottom has some specific characteristics that can be found in the article about deep sea bottom.

Adaptations

For the animals of the deeper layers of the water column where there is no plant growth at all, finding food and avoiding predation are the main problems. Many mid-water animals have bizarre body shapes, because they never come into contact with a physical barrier. The most abundant for is more or less spherical, often with long and delicate extensions. Most animals do not need a rigid skeleton because they spend all their time suspended in water. An important adaptation is not to be seen too easily. In the upper parts of the mesopelagic zone, where there is still quite some sunlight, many animals have more or less transparent bodies. An alternative is to have silver sides that act as mirrors and light organs along their bellies. At the deeper layers, the organisms have a dark red, black or brown colour. These colours do not reflect the blue colour of most bioluminescence. In the mid-water, animals have mouths full of big teeth. In the deeper water, the anglerfish has a fishing rod with bioluminescent lures to attract preys. Many fishes also have large mouths to take up as much as possible. Another adaptation is dial vertical migration of many animals. The general pattern is that animals spend the daylight hours at depth and move upwards during the night. This is to avoid predation and to feed. Bioluminescence is another adaptation. Depending on the depth, organisms use it as camouflage to protect them from predators or to confuse or frighten predators rather than to hide from them. Other animals use it to look for a potential mate. A few deep-sea animals use it together with another light organ that produces deep red light. Other animals cannot see this red light, but the red light will be reflected very efficiently by the red bodies of the specific prey. The blue light is a recognition mark for other members of the same species. An example of such an organism is *Malacosteus*.

Biota

The mid-water communities include representatives of almost every major animal group on Earth, from tiny single-celled organisms only a fraction of a millimetre, through all the invertebrate phyla, to fish several meters long. Most of them have an average size in the range of centimetres. The dominant species are crustaceans, fish and a variety of animals with soft and jelly-like bodies such as jellyfish.

Squid with many light organs

On the deep-sea floor, many mounds and depressions are formed by benthic animals such as worms, mollusks, crustaceans, starfish, brittlestars, shrimps, fishes sea cucumbers and sea urchins. Because the deep-sea floor is poor in food resources, it cannot support large populations of big animals. This causes bigger animals to be distributed over a greater range than elsewhere. For this reason, many animals adopt a 'sit and wait' strategy. This strategy means that the animals stay at one place and wait until food falls down on them or are carries to them in the currents. These animals are sponges, sea anemones, tube worms and barnacles. Smaller animals can also be found.

Deep-sea bottom with fish

A gorgonian feeding on a basketstar

There are five groups that are dominated in terms of numbers. These are: foraminiferans, Radiolaria, nematodes, crustaceans (amphipod shrimps, isopods, copepods, bivalves) and polychaete worms.

Foraminifera

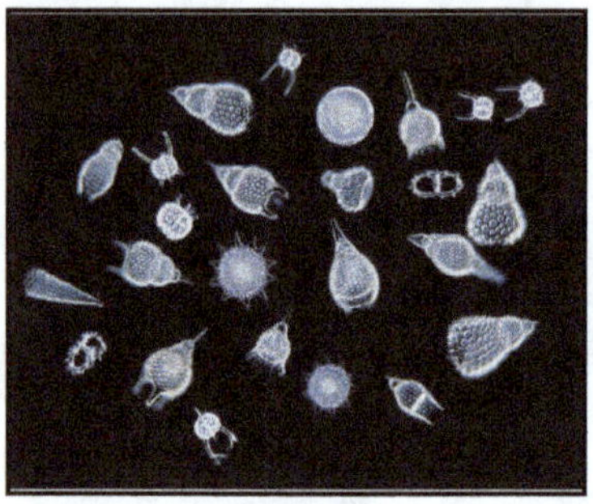

Radiolaria

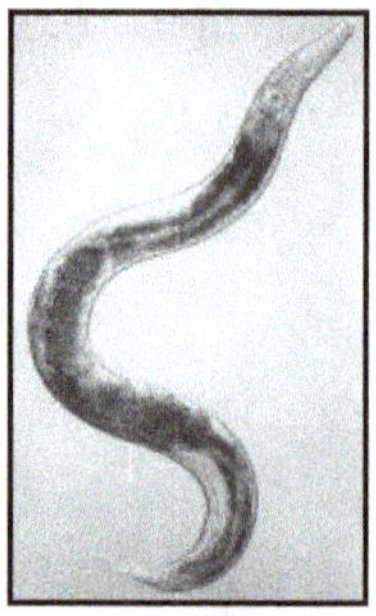

Nematode

Isopod

Polychaete worm

Riftia with associated animals

Hydrothermal communities may be hundreds of times more abundant than on the adjacent sea floor. This is because the animals have their own rich source of food. This food source is totally independent of the input from the photosynthesizing plants in the overlying surface waters. Organisms that can be found in association with these vents are bacteria, mussels, cockles, oysters, sea anemones, shrimps, crabs, fish and tube worms. An example of a tube worm is *Riftia pachyptila*. It has a red plume of gills sticking out of the top of the worm's white tube. They are also found around cold hydrocarbon seeps. It has no mouth or gut and it has special chemosynthetic bacteria in the tissues of its body called the trophosome. The chemicals that the bacteria need are delivered by the blood of the worm.

Cold seeps have a lower metazoan biodiversity but a higher productivity. This is because of the high biomass of megabenthic organisms. The communities are often dominated by the bivalves that have symbiontic relationships with chemosynthetic bacteria. Organisms that can be found here are tube worms, bivalves, crabs, anemones and soft corals. An example of an organism is the *vesicomyid* clam that has a symbiontic relationship with sulfur oxidizing bacteria.

Seamounts have high concentrations of plankton. Also fishes are abundant, because of the high concentration of plankton and the constant influx of prey organisms. Marine mammals such as sharks, tuna and cephalopods all congregate over the seamounts to fee. Even seabirds have been shown to be more abundant in the zones with shallow seamounts.

Carcasses are hot-spots for food. When the carcass reaches the bottom, it attracts within minutes amphipods. They have powerful jaws, which can slice flesh. They can detect the odor of a carcass carried on the water currents. After them several fishes arrive and feed on the carcass. Also shrimps can occur on it. When these animals have eaten enough or when there is nothing suitable left for them, the remainders will attract slow-moving animals such as mollusks, starfish, brittlestars and sea cucumbers. Finally, bacteria will finish the job. They can even break down the bones. Within weeks the carcass is cleared away.

Giant animals and strange creatures are common in the deep sea. These are sometimes called monsters of the sea. There are a lot of examples. Many myths of monsters must be based on sightings of squid. Sharks are also seen as monster and aggressive animals. Other animals can be seen as monsters because they are very large in comparison to the species that we know. An example is the giant isopod. This phenomenon is called gigantism.

Shark

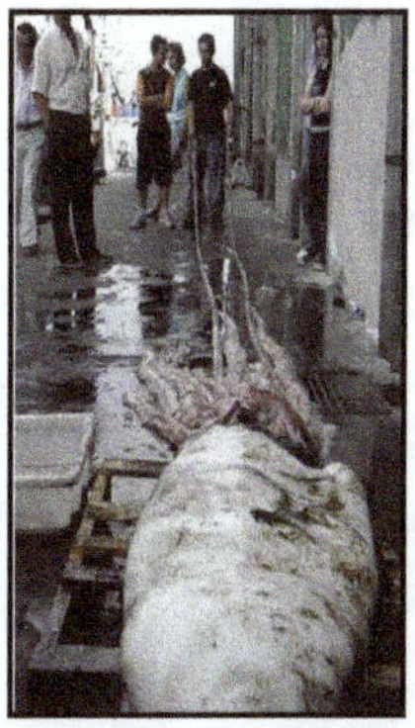

Giant squid

Giant isopod

Sea spider

Threats

Human activities can cause serious damage to the deep sea. Examples of human activities that have a negative impact are:

Fishing activities: overfishing, destructive practices and illegal, unreported and unregulated fishing activities (IUU);

- Bycath or incidental mortality of non-target species;
- Noise pollution such as powerful sonar systems and airguns;
- Shipping: noise, accidental spills of oil, discharge of garbage, oily wastes, sewage, chemical residues and ballast water, anti-fouling products;
- Submarine cables and pipelines;
- Pollution;
- Climate change and ozone depletion.

Deep-sea Ecosystems: Environmental Characteristics and Biodiversity

The oceans cover 70% of the Earth's surface or the equivalent of the surface of two Mars and two Moons together. But we still know more about the geography and characteristics of our Moon or Mars than about our Oceans! Furthermore, 50% of the Earth is covered by oceans more than 3,000 m deep, with a mean depth of around 3,800 m. The deep sea is, therefore, the largest ecosystem in our planet as well as one of the least studied. It comprises a variety of habitats from the shelf break to the deepest parts of the ocean floor found in trenches, each with specific physical and geochemical characteristics that support one of the highest biodiversities on the planet. In relation to the energy that supports marine ecosystems, deep-sea habitats can be divided into two major groups: heterotrophic and chemosynthetic habitats. In heterotrophic habitats, the faunal communities depend, ultimately, on organic matter produced at the surface by photosynthesis and are therefore dependent on solar energy. In chemosynthetic habitats, the biological communities are sustained by the energy provided by inorganic reduced chemicals such as hydrogen sulphide (H_2S) or methane (CH_4) from the Earth's interior.

Heterotrophic Ecosystems

The vast majority of life in the deep oceans is sustained by the production of organic matter on the surface from photosynthesis. It resides in what are known as heterotrophic habitats, because there is no intrinsic primary production. In the deep-sea benthos, the heterotrophic ecosystems include continental margins from the shelf break to 3,000 m depth and abyssal plains, between 3,000 m and 6,000 m in depth.

Sediment Margins

Continental margins cover 13% of the world's seafloor. These systems are the largest reservoir of sediments on Earth, with up to 90% of sediments generated by erosion on land being deposited on the margins. The open margin ecosystem is greatly influenced by dynamic processes such as currents that affect and drive the transport of energy and organic matter. In some regions, wind stress along the coast can lead to upwelling events that transfer rich deep waters to the surface, feeding the nutrient depleted surface waters and resulting in high productivity on the shelf. In other areas, landslides cause large-scale disturbances that can destroy whole communities in a single event.

The drivers of heterogeneity in faunal distribution, composition and abundance on continental margins vary depending on the spatial scale considered. At large scales (over 1,000 km), major physical factors such as geology, temperature, currents and water masses play the main role. At mid scales (1-100 km), the distribution of animals is mainly determined by factors such as down flux of primary production, oxygen availability (i.e., areas of oxygen minima), sediment type and catastrophic events. Finally, biological interactions are the main drivers of faunal distribution at small scales. Our knowledge about the biodiversity and biogeography of fauna on continental margins is still scant. Biomass as indicated by epifauna (animals living on the sediment) decreases with depth, and in deep waters the presence of a large number of burrowing animals is shown by a variety of features such as pits and mounds. Animals on deep sediment slopes are mainly sediment feeders that use the organic matter input from the surface. In shallower waters, the number of megafaunal animals and suspension feeders increases in relation with higher water currents. Finally, when the slope approaches the shelf, the increase in grain size causes a decrease in the biota of the sediment. One of the most striking observations in open margins is the peak in biodiversity at mid slopes. The exact drivers of this general observation of biodiversity maxima are still to be determined, and are the focus of a number of research projects.

Canyons

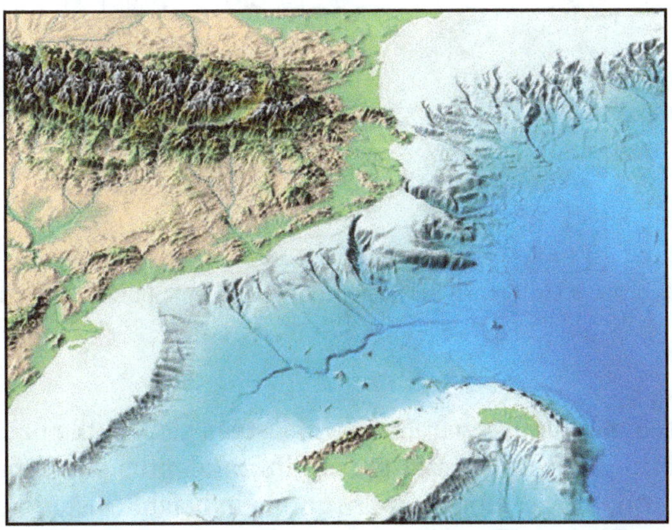

Bathymatric map of a section of catralan sea (eastern Mediterranean) showing canyon systems

Canyons are deep incisions on the continental margins, and are common features on European margins such as the Catalan Sea, the south of France or the Portuguese margin Canyons are hotspot ecosystems on continental margins, characterised by a high biodiversity. These geological features act as major pathways for organic carbon transportation, and fast down flux of organic matter from the land to the deep sea. Canyons contain a variety of substrata, such as hard rock walls and mobile sediments on the canyon floor that sustain complex ecosystems with a high degree of endemic species. Canyons are also important hotspots for commercial species, such as the red shrimp Aristeus antennatus, one of the major fisheries in the Catalan Sea. However, their irregular topography and the difficulty of sampling prevented their detailed investigation until only recently. The latest developments in deep-water imaging with towed and remote-operated vehicles and submersibles are now facilitating the exploration and investigation of the geophysical and biological characteristics of canyons.

Deep-water Corals

Investigations on continental margins during the last decade led to a surprising discovery: the presence of deep-water corals that form reefs along the NE and NW Atlantic continental margins. The NE Atlantic coral reefs are found at around 1,000 m depth and extend from Norway to Portugal, and recently similar ecosystems have been discovered in the Mediterranean. Deep-water coral species such as Lophelia pertusa and Madrepora oculata form carbonate reefs several kilometres in length and sustain a high biodiversity providing refuge, structure and nursery spots for other slope species. The reefs provide a complex three-dimensional habitat for a variety of species, including sponges, soft corals, molluscs, crustaceans and echinoderms, as well as for commercial species. Although our knowledge on the composition and functioning of these rich communities is still low, there is already evidence of habitat damage from deep-water trawling over deep-water coral regions.

Deep-sea corals observed by French ROV Victor2000 at a depth of 1,650 m in the NE Atlantic

Seamounts

Seamounts are undersea mountains characterised by steep slopes, the presence of hard and soft substrata, large depth ranges from abyssal to sub-littoral depths and geographic isolation. It is estimated that around 100,000 seamounts over 1000 m in height exist around the world's oceans, and many more if we consider smaller mounts. But only around 350 of these seamounts have been sampled, and only around 100 have been studied in any detail. The particular biological features of seamounts include high productivity, large stocks of commercially valuable fishes, high biodiversity and a high degree of endemism of benthic fauna. These specific traits are driven by the particular topography and hydrography around seamounts.

As has occured in other regions, like canyons, with difficult terrain, we still have little knowledge of the biodiversity, distribution and functioning of seamount fauna. However, seamounts have been the target of intensive fishing in recent decades, which has led to potential long-term damage and biodiversity loss in an ecosystem as yet poorly understood. Today, with the help of new studies of seamounts driven by the use of new technologies such as ROVs or deep-towed cameras, management and conservation options are being put in place.

Anoxic Areas

Mid-water oxygen minima (<0.5ml/l dissolved O_2) can intercept the continental margin, resulting in sediments with a very low oxygen concentration or Oxygen Minimum Zones (OMZs). OMZs are formed in areas of high primary production in the surface waters of the ocean and poor water circulation, where the biological degradation of the sinking organic matter results in oxygen depletion. Seafloor OMZ's typically occur between 200 m and 1000 m depth and are found in the eastern Pacific, NW Pacific margin, Philippines area, Bay of Bengal, Arabian Sea and SW Africa beneath the Benguela current. Despite very low oxygen concentrations, protozoan and metazoan life thrive in these ecosystems. The high concentrations of organic matter sustain dense populations of sulphide-oxidising bacteria (i.e., Begiattoa, Thioploca, Thiomargarita) and a low biodiversity but high density of protozoan and metazoan life. The main groups are foraminiferans, nematodes, ciliates, flagellates, polychaetes, gastropods and bivalves with specific adaptations, such us high concentrations of haemoglobins, large respiratory surfaces, small thin bodies, high concentrations of pyruvate oxydoreductases and presence of sulphide-oxidising symbionts.

Abyssal Plains

The abyssal plain ecosystem is the largest ecosystem on Earth. It lies beyond the continental slope, between 3,000 and 6,000 m depth. Abyssal plains are covered by a thick layer of fine sediment that can reach thousands of metres in thickness, resulting in the popular picture of a flat, monotonous deep-sea bed. The main characteristics of water masses at abyssal plains are: low temperature (~2 °C except in the Mediterranean Sea with 13 °C and Red Sea with 21.5 °C), salinity (35‰, except in the Mediterranean and Red Sea >39‰), mostly saturated waters with dissolved oxygen (5-6 ml/l), absence of light (light useful for photosynthesis does not reach below ~250 m depth) and high pressure (1 atmosphere every 10 m depth). This relatively uniform distribution of physical factors led to the belief that abyssal plains were very stable habitats where physical and biological processes remained unchanged over short and long time scales.

There is now evidence that physical disturbances occur at abyssal plains, causing important biological responses. For example, there are daily and annual tidal variations in the flow of cold dense water close to the seafloor. The effects of these tides on the biological communities are not well understood, but it has been suggested that they could be used by certain species for orientation or for setting internal biological cues for synchronised spawning. There are also high-energy, unpredictable events such as benthic storms or turbidity currents that have very considerable disruptive effects on the seafloor, in particular in the redistribution of sediment and consequent biological responses. Another major environmental factor that greatly affects the benthic communities on abyssal plains is the seasonal deposition of phytodetritus (organic matter produced in the surface waters) following the months of high surface production. Because the rapid sinking of this material prevents its complete utilisation by pelagic grazers, the arrival of this organic matter to the seafloor provides the abyssal communities with a seasonal input of high-quality food resource.

The abyssal plains support a very high biodiversity, composed mainly of macro and meiofauna. The meiofauna (size of organisms in the order of microns) is mainly dominated by nematodes and foraminifera. The macrofauna (size of organisms in the order of millimetres) is dominated by polychaetes, with small peracarid crustaceans, molluscs, nemerteans, sipunculans, echiurans and enteropneusts also abundant. Finally, the large megafauna (size of organisms in the order of centimetres) is made up of holothurians, asteroids, echinoids, decapod crustaceans and fish, as well as sessile fauna such as crinoids, sponges and anthozoans on hard substratum.

Even though abyssal plains have been sampled since the times of the Challenger expedition, only a small fraction of the vast extensions of these ecosystems has been studied to date. Latest results obtained from abyssal plain research have shown that variations in primary production in the surface waters can result in long-term changes in the composition of the plain megafauna. For example, there is evidence from the Porcupine Abyssal Plain in the NE Atlantic that an almost non-existent species of small holothurian (Amperima rosea) became dominant after 1996 because of its ability to rapidly exploit the nutritional resources of seasonal phytodetritus. This indicates the strong link between the abyssal ecosystem and the surface of the biosphere, and has important consequences when considering the effect of factors such as climate change on biodiversity.

Chemosynthetic Ecosystems

Deep-water chemosynthetic ecosystems have been known and studied for less than 30 years. The first such ecosystems to be discovered were hydrothermal vents in 1977 8 years after Neil Armstrong and Buzz Aldrin had walked on the Moon! Then followed the discoveries of other deep-water chemically driven communities such as cold seeps, large organic falls to the deep-sea floor (i.e., whale falls or sunken wood and kelp) and areas of oxygen minimum that intersect with the margin. In chemosynthetic ecosystems, primary production is produced by chemoautotrophic microorganisms that use reduced inorganic chemicals to synthesise organic matter. These organisms are found free living, forming bacterial mats, but also in symbiosis with some of the major invertebrate groups.

Hydrothermal Vents

Hydrothermal vents were discovered in 1977 in the Galapagos Rift, in the Pacific, and since then vents have been found in all ocean basins. Hydrothermal vents occur in mid-ocean ridges, back-arc basins and certain active seamounts. Mid-ocean ridges are volcanic mountain chains that occur where two tectonic plates are being pulled apart. In these areas, cold seawater (2 °C) penetrates through cracks in the crust. During its transition in the mantle, the fluid gets heated as it flows close to the magma chamber that feeds the ridge and is depleted of oxygen and magnesium while being charged with other metals. The superheated fluid (350 °C) raises back to the surface of the seafloor, and when it mixes with the surrounding cold and oxygenated seawater, the metals precipitate, providing the aspect of dense black smoke characteristic of hydrothermal vents.

Among the most striking discoveries at vents were the associated dense biological populations and the trophic structure that sustains these communities. It was unforeseen to find whole dense communities of animals living independently from solar energy by using the energy of reduced chemicals from the Earth's interior via the production of microorganisms. But it was even more astonishing to find that these microorganisms also formed symbiotic relationships with most of the major invertebrate groups, and to observe the variety of adaptations made by these invertebrates to life in hydrothermal vents. One of the most modified is probably the giant tubeworm Riftia pachyptila from the Pacific vents. This animal does not have a mouth or digestive system, but instead has a special organ that fills most of its body, called the trophosome. The trophosome is basically a sack densely packed with chemoautotrophic bacteria. Riftia intakes oxygen from the surrounding water and CO_2 and H_2S from the hydrothermal fluid with its highly irrigated plume. The chemicals are sent to the trophosome via the blood vessels where the microorganisms use them to synthesise organic matter. The animal depends completely on this microbial production for its lifelong growth and reproduction. Symbiotic relationships also appear in other groups, such as clams, mussels, shrimp, crabs and polychaetes, with different degrees of dependency.

Black smoker from the Mid-Atlantic Ridge

Gastropods from the hydrothermal vents of the Lau Basin, in the western Pacific

Hydrothermal vents have been called "oases" of life in the deep-sea floor because of the exuberant aspect of their dense populations of large invertebrates. However, as in other ecosystems with extreme chemicophysical environmental parameters, hydrothermal vent communities are simple systems. Biodiversity is low, but biomass is high, sustained by a constant and abundant supply of energy in the form of reduced chemicals found in the hydrothermal fluids. Since their discovery in 1977, 590 species have been described from vents, which is the equivalent to around one new description every two weeks. Furthermore, of the almost 600 species described, approximately 400 have been identified so far as endemic to vents. The major faunal groups present are vestimentiferan tubeworms, bathymodiolid mussels, vesicomyid clams, bresilid shrimp, crabs, amphipods and polychaetes. Investigations at hydrothermal vents are still in the extensive exploration phase, with only a small fraction of the over 65,000 km of global ridge system studied to date. However, the data that has been compiled so far indicates that vent fauna form distinct biogeographical regions.

In a review by Van Dover et al., six vent biogeographic regions are recognised, each with specific faunal assemblages. But much exploration and investigation remains to be done before we have a sound understanding of the global diversity of vent species, and the processes that shape their distribution and their functioning.

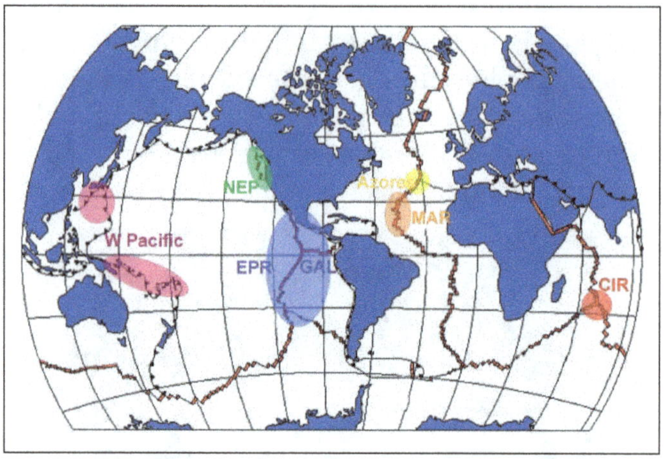

The mid-ocean ridge system showing the known hydrothermal vent biogeographic provinces

Hydrothermal vent biogeographic provinces. Azores: dominated by bathymodiolid mussels, amphipods and caridean shrimp; MAR: Northern MidAtlantic Ridge region dominated bycaridean shrimp,mainly Rimicaris exoculata, and bathymodiolid mussels; EPR &GAL: East Pacific Rise and Galapagos Rift dominated by vestimentiferan tubeworms, bathymodiolid mussels, vesicomyid clams, alvinellid polychaetes, amphipods and crabs. NEP: NE Pacific region, dominated by vestimentiferan tubeworms excluding Riftiidae, polychaetes and gastropods; W Pacific: dominated bybathymodiolid mussels, "hairy" gastropod, vesicomyid clams and shrimps; and CIR: Central Indian Ridge, dominated by the shrimp Rimicaris kairei, mussels, scale gastropods and anemones.

Cold Seeps

A bathymodiolid mussel community in Gulf of Mexico cold seeps

Cold seep communities were discovered in 1983 at aproximately 500 m depth in the Western Florida Escarpment in the Gulf of Mexico. Cold seeps are characterised by the seepage of cold fluid with a high concentration of methane. This methane may have a biological origin, from the decomposition of organic matter by microbial activity in anoxic sediments, or a thermogenic origin, from the fast transformation of organic matter caused by high temperatures. Cold seeps also have high concentrations of H_2S in sediments, produced by the bacterial reduction of sulphates using methane. Both methane and sulphide play a major role in sustaining the highly productive cold seep communities through chemoautotrophy by free-living and symbiotic bacteria. Cold seep communities occur in both passive margins such as the Gulf of Mexico, Carolina slope, Barents Sea, Gulf of Guinea and Angola margin, and in active margins (or subduction zones), mainly in the Pacific, such as the Peru-Chile margin, as well as the Barbados Accretionary Prism and the Eastern Mediterranean among others.

Tubeworms of the genus Lamellibrachia from Gulf of Mexico cold seeps

As with hydrothermal vents, only a small fraction of the potential locations of cold seeps on margins has been explored to date.

Since their discovery, around 230 species have been described from cold seeps. Cold seep habitats are more stable systems than hydrothermal vents. There is also a slow transition of physical and chemical factors between the seep habitat and the heterotrophic surrounding system, allowing for a higher biodiversity than in hydrothermal vents. The megafaunal biomass at seeps by far exceeds that of the surrounding non-chemosynthetic sediment. The major groups are bivalves (mytilids, vesicomyids, lucinids and thyasirids) and vestimentiferan tubeworms, with pogonophoran, sponges, gastropods and shrimps sometimes also abundant.

Other Reducing Habitats

In 1987, Craig Smith, from the University of Hawaii, observed for the first time chemosynthetic communities on a whale skeleton that was found by chance in the North Pacific during a dive with the submersible Alvin.

Since then, the investigation of biological assemblages on whale falls and other large organic falls to the deep-sea floor, such as sunken wood and kelp, has advanced rapidly.In the case of whale falls, thereis a three-step ecological progression. First, during the scavenger phase, the flesh is eaten and the skeleton left exposed. The opportunistic phase follows, when the sediment and skeleton are colonised by dense populations of opportunistic polychaetes and crustaceans. The final phase is the chemotrophic or sulphophilic phase. The bones of whales are composed 60% of lipids. The anaerobic bacterial degradation of these lipids produces sulphides that are used by chemoautotrophic microorganisms, allowing for the subsequent colonisation of chemically-driven fauna.

Whale skeleton colonised by bacterial mats

The biodiversity of fauna colonising these isolated and ephemeral habitats is high. Since their discovery, over 400 morphological species have been described from whale falls, but most of them still remain to be identified. It has been suggested that whale falls could act as stepping stones for dispersal between chemosynthetic ecosystems. This is supported by the fact that the three habitats share a number of species and an even higher number of groups at higher taxonomic levels.

Finally, chemosynthetic-related communities can also develop in OMZ's. A large number of heterotrophs in OMZs consume chemoautotrophic bacteria by grazing on bacterial mats or predating on other animals that have done so. The presence of endosymbiotic sulphur-oxidising bacteria is also widespread in foraminifera's, flagellates, ciliates, some polychaetes and some bivalves. The details of the metabolic interactions between host and symbiont remain unknown, as does the extent to which chemosynthesis provides nutrients to the OMZ benthos. But ongoing and future research will no doubt extend the list of these types of relationships, and help explain the phylogenetic and evolutionary links with fauna from other deep-water chemosynthetic ecosystems.

Of all the described species from hydrothermal vents, cold seeps and whale falls, 18 are shared between vents and seeps, 11 are shared between vents and whales, 20 are shared between seeps and whales, and 7 are shared amongst the three habitats. However, these numbers will change in parallel with new discoveries and further investigation of known sites that will improve our knowledge of the diversity and distribution of species from deep-water chemosynthetic habitats and the processes driving them.

References

- Marine-ecosystem, science: britannica.com, Retrieved 30 March, 2019
- Threats-to-coral-reefs, coral-reef, science: britannica.com, Retrieved 3 July, 2019
- Coral-reef: newworldencyclopedia.org, Retrieved 22 May, 2019
- Mangrove-Ecosystems: aquafind.com, Retrieved 21 April, 2019
- What-is-a-mangrove: ecologic.org, Retrieved 25 June, 2019
- Ecosystem-of-the-Mangroves-Brandon-Wolf, forestry: uwsp.edu, Retrieved 10 August, 2019

- Kelp-Forest-Ecosystems, Marine-Environments, Oceanography: libretexts.org, Retrieved 14 May, 2019
- Rocky-shore, marine-habitats, coasts-waterways, environment: qld.gov.au, Retrieved 30 January, 2019
- Rocky-shore, marine, environment: basicbiology.net, Retrieved 2 March, 2019
- Threats-to-sandy-beach-ecosystems: researchgate.net, Retrieved 4 June, 2019
- Saltmarsh: oceanservice.noaa.gov, Retrieved 15 February, 2019
- Salt-marsh-Characteristics: newworldencyclopedia.org, Retrieved 5 April, 2019
- Salt-marsh-Importance: newworldencyclopedia.org, Retrieved 3 July, 2019

Permissions

All chapters in this book are published with permission under the Creative Commons Attribution Share Alike License or equivalent. Every chapter published in this book has been scrutinized by our experts. Their significance has been extensively debated. The topics covered herein carry significant information for a comprehensive understanding. They may even be implemented as practical applications or may be referred to as a beginning point for further studies.

We would like to thank the editorial team for lending their expertise to make the book truly unique. They have played a crucial role in the development of this book. Without their invaluable contributions this book wouldn't have been possible. They have made vital efforts to compile up to date information on the varied aspects of this subject to make this book a valuable addition to the collection of many professionals and students.

This book was conceptualized with the vision of imparting up-to-date and integrated information in this field. To ensure the same, a matchless editorial board was set up. Every individual on the board went through rigorous rounds of assessment to prove their worth. After which they invested a large part of their time researching and compiling the most relevant data for our readers.

The editorial board has been involved in producing this book since its inception. They have spent rigorous hours researching and exploring the diverse topics which have resulted in the successful publishing of this book. They have passed on their knowledge of decades through this book. To expedite this challenging task, the publisher supported the team at every step. A small team of assistant editors was also appointed to further simplify the editing procedure and attain best results for the readers.

Apart from the editorial board, the designing team has also invested a significant amount of their time in understanding the subject and creating the most relevant covers. They scrutinized every image to scout for the most suitable representation of the subject and create an appropriate cover for the book.

The publishing team has been an ardent support to the editorial, designing and production team. Their endless efforts to recruit the best for this project, has resulted in the accomplishment of this book. They are a veteran in the field of academics and their pool of knowledge is as vast as their experience in printing. Their expertise and guidance has proved useful at every step. Their uncompromising quality standards have made this book an exceptional effort. Their encouragement from time to time has been an inspiration for everyone.

The publisher and the editorial board hope that this book will prove to be a valuable piece of knowledge for students, practitioners and scholars across the globe.

Index

A

Aethia Pusilla, 44

Aethiini, 44, 46

Aipysurus Eydouxii, 30

Albatrosses, 42-43, 50-60

Anaerobic Metabolism, 18, 20

Aptenodytes Forsteri, 63, 65-66

Auk, 44-45, 72

B

Beluga Whales, 131

Benthic Zone, 10

Billfish, 16

Biotic Factors, 4

Black Sea, 12, 192, 200

Blue Marlin, 13-14, 16

Bluefin Tuna, 11-12, 112, 138, 143

C

Caretta Caretta, 22-23, 25

Cerebral Cortex, 38

Chelonia Mydas, 21-25

Chelonioidea, 21-22, 24

Chicks Chirrup, 48

Ciliates, 7-8, 86, 135, 137, 139, 207, 212

Comb Jellies, 4, 137

Common Fangtooths, 14-15

Continental Shelf, 6, 9-10, 44, 133-134, 139, 145, 152-153, 199

Coral Reefs, 1, 5-7, 17, 31, 75, 79, 133, 135, 140-141, 143, 145-148, 150-162, 164-165, 167, 206

Crustaceans, 14, 16-17, 25, 30, 32-33, 40, 54, 114, 136-140, 147, 156, 158-159, 164, 168, 176, 182, 185, 194, 201-202, 206, 208, 211

Cyanobacteria, 7, 81-84, 87-88, 95-96,132, 157, 178

D

Deep-sea Vents, 140, 144

Dermochelys Coriacea, 20-23

Disphotic Zones, 133

Dorsal Fins, 16

Dugongidae, 128-130

Dugongs, 101, 105, 123, 128-130

E

Echinoderms, 132, 146-147, 158, 206

Enhydra Lutris, 127, 173

Enhydris Bennettii, 32

F

Fiord, 9

Fordonia Leucobalia, 32-33

Fraterculini, 44, 46

G

Gharials, 35, 37

Guineafowl Puffer, 15

H

Hind Toe, 50, 52

Hydrobates Pelagicus, 50

I

Ice Cover, 48, 50

J

John Dory, 15-16

K

Krill, 42, 44-45, 49, 51, 54, 56, 62, 77, 94, 112, 125, 135, 137-138

L

Leopard Seal, 62-63, 125

Lepidochelys Kempii, 22-23

Longline Fishing, 43

M

Marine Biota, 136

Marine Birds, 42

Marine Habitat, 1, 8

Marine Iguana, 11, 18-21, 34-35

Marine Microbiology, 2

Marine Reptiles, 11, 18-20

Mitochondrial Dna, 60

Molluscs, 9, 90, 92, 97, 178, 182, 185, 194, 206, 208

Mollymawk, 60-61

Mustelidae, 101, 126-128, 131

N

Nekton, 25, 93, 136, 138-139, 142, 144

Nematocysts, 4, 137

Nutrient Supply, 7

O

Ocean Floor, 1, 134, 138, 150-151, 199-200, 204

Orcas, 16, 62, 102

P

Pelamis Platurus, 20, 29

Petrel, 46-50, 74

Phocid Breeding, 125

Photic Zone, 6, 133-134, 136, 139, 142

Procyonidae, 126-127

Protozoa, 7-8, 75, 86-87, 132, 144

S

Scorpionfish, 17-18

Sea Anemones, 4, 141, 159, 201-202

Sea Otters, 4, 101-102, 104, 123, 125, 129-130, 173

Sea Slugs, 4, 9, 159

Sea Squirts, 9, 136, 158, 178

Sirenia, 105, 128-129

Skunks, 123, 126-128

Soft Corals, 9, 15, 179, 203, 206

Sooty Albatrosses, 51-52, 55, 60

Spheniscidae, 61, 65, 67-68

Squid, 16, 42, 47, 49, 51, 54-56, 62, 103, 110, 112, 125, 136, 139, 142, 144, 160, 201, 203

T

Thalassarche, 51, 59-61

Z

Zooplankton, 3, 11, 13, 16, 25, 54, 80, 90, 94-96, 119, 137-138, 144-145, 156-157, 182

www.ingramcontent.com/pod-product-compliance
Lightning Source LLC
Jackson TN
JSHW061531060426
100415JS00005B/30
* 9 7 8 1 6 4 1 7 2 6 6 7 2 *